Ecological Indicators for Coastal and Estuarine Environmental Assessment

WITPress

WIT Press publishes leading books in Science and Technology.
Visit our website for the current list of titles.
www.witpress.com

WITeLibrary

Home of the Transactions of the Wessex Institute, the WIT electronic-library
provides the international scientific community with immediate and permanent
access to individual papers presented at WIT conferences. Visit the WIT eLibrary
at http://library.witpress.com

Ecological Indicators for Coastal and Estuarine Environmental Assessment

A user guide

J.C. Marques
F. Salas
J. Patrício
H. Teixeira
J.M. Neto

· U   C ·

UNIVERSIDADE DE COIMBRA

WITPRESS Southampton, Boston

J.C. Marques, J. Patrício, H. Teixeira & J.M. Neto
IMAR, Institute of Marine Research, University of Coimbra, Portugal

F. Salas
Univerisity of Murcia, Spain

Published by

WIT Press
Ashurst Lodge, Ashurst, Southampton, SO40 7AA, UK
Tel: 44 (0) 238 029 3223; Fax: 44 (0) 238 029 2853
E-Mail: witpress@witpress.com
http://www.witpress.com

For USA, Canada and Mexico

WIT Press
25 Bridge Street, Billerica, MA 01821, USA
Tel: 978 667 5841; Fax: 978 667 7582
E-Mail: infousa@witpress.com
http://www.witpress.com

British Library Cataloguing-in-Publication Data
A Catalogue record for this book is available
from the British Library

ISBN: 978-1-84564-209-9

Library of Congress Catalog Card Number: 2009920915

*The texts of the papers in this volume were set
individually by the authors or under their supervision.*

Biographies

João Carlos Marques
Born in Lisbon (Portugal) in April 1957, he graduated in Biology in 1980 at the University of Lisbon, specializing in Marine Biology. He concluded his PhD in Ecology in 1989. He is, since 2003, Full Professor at the Faculty of Sciences and Technology, University of Coimbra (FCTUC). He was President of the Scientific Board of the Department of Zoology from FCTUC between 1999 and 2003, and Vice-Rector for Scientific Research of the University of Coimbra from 2003 to 2007.

He has led the IMAR – Institute of Marine Research Consortium in Portugal between 1997 and 2006. The IMAR Consortium was founded in 1991, associating presently approximately 300 researchers from several R&D Units. Besides leading the IMAR Consortium as a whole, he has been the scientific Coordinator of IMAR – Marine and Environment Research Centre, one of IMAR's R&D Units.

A strong interest in biological and ecological processes in marine and estuarine ecosystems, system's ecology and the analysis of ecosystem integrity and ecological quality status has been at the core of his teaching and research activities. He has coordinated several large research projects, namely under the aegis of European Union programmes. In this scope, since 1989, he supervised the research of a large number of MSc and PhD students, in Portugal and abroad. So far, he authored or co-authored hundred and thirty scientific papers in international refereed journals, as well as several books, books chapters and international proceedings. He is a member of the Editorial Board and/or Associate Editor of several international research journals, both in Europe and USA.

Fuensanta Salas

Born at Murcia (Spain) in 1969, she graduated as Biologist at the University of Murcia, where she also obtained her PhD in 2002. Her doctoral dissertation focused on the evaluation of the applicability of different indicators of organic enrichment in the management of coastal marine environments.

From 2002 to 2004 she was a Post Doctoral researcher at IMAR – Institute of Marine Research, at the Coimbra University, Portugal. She participated in several scientific research projects, namely in the scope of European Union Programmes, and authored or co-authored significant papers on the development and validation of ecological indicators, published in international refereed journals and in books chapters. At present, she continues collaborating actively at IMAR. Her research interests include the study of marine and estuarine ecosystems and marine protected areas, benthic ecology, ecological indicators and environmental management.

Joana Patrício

Born in Coimbra (Portugal) in 1977, she finished her PhD in Biology at the University of Coimbra in 2005. Her thesis analysed how well ecological indicators assess environmental status, based on a number of cases studies from coastal and estuarine ecosystems.

Among her recently published works one may find titles on theoretical ecology, ecological indicators performance, network analysis and modelling. She holds presently a Researcher position at IMAR – Institute of Marine Research, at the Coimbra University. Her research has been contributing to a more soundly based political decision with regard to coastal and transitional waters ecosystems environmental quality assessment and management. Moreover, since 2003, she has been actively involved as invited teacher in under and post graduate course in Marine Biology and Ecology, application of ecological indicators, and ecological quality assessment, both in Portugal and abroad. She is presently enthusiastically involved in the implementation of the EU Water Framework Directive (WFD) in Portugal, as well as in several European research projects in relation to the sustainable use and management of Mediterranean freshwater, transitional, and coastal water bodies, through a socioeconomic and environmental analysis of changes and trends to enhance and sustain stakeholders' benefits.

João Neto

He was born in Mozambique in 1968. His academic background as a biologist began at the University of Coimbra (Portugal), through areas as diverse as studies concerning macroinvertebrates communities, nutrients mass balance in estuaries and seagrass meadows and salt-marsh dynamics. Those different pieces of information allowed him to increase his knowledge in ecological modelling, eutrophication processes and restoration and management of estuarine systems, and were the basis of his PhD at the same University. His participation in several international projects, and cooperation with foreign institutions and Universities, allowed him to increase his management knowledge and experience in catchment basins and coastal systems. Presently, he holds a Researcher position at IMAR – Institute of Marine Research at the University of Coimbra, which is mainly focused in coastal and estuarine ecology. The research interests are now focused in coastal and transitional waters quality assessments, namely in the scope of the implementation of the EU WFD in Portugal. As a feedback to recent demands from European water policies, the assessment methodologies for different biological quality elements stated in the WFD are under development, as well as the integration methodologies for the classification of the total water bodies' quality status. Transport processes and the influence of bio-turbation versus sediment consolidation in eutrophication processes in estuaries, and integrative management of coastal and transitional waters systems are also part of his main research interests.

Heliana Teixeira

Born in Guimarães (Portugal) in 1978, she graduated in Biology in 2001 at the University of Coimbra, where she also obtained her MSc degree in Ecology in 2005. She is presently a PhD student in Ecology, and her research focuses on the development and improvement of ecological indicators for ecosystems' ecological quality evaluation and management.

She is presently member of the IMAR – Institute of Marine Research team at the Coimbra University, and has been collaborating on several research projects related to Portuguese ecosystems' recovery and management, and to the implementation of the European WFD in coastal and transitional waters. Moreover, in the scope of her PhD work she has been working in close cooperation with researchers from other institutions in Europe and in USA, which contributed to improve her training on the use of benthic ecological indicators. So far, she has authored or co-authored several significant papers in the field, published in international refereed journals.

Contents

Preface

Ecological indicators are commonly used to provide synoptic information about the state of ecosystems, and when applied effectively are expected to reveal conditions and trends that will help in development planning and decision making processes. Nevertheless, the application of ecological indicators is not exempt from criticism, and therefore should be handled following the right criteria and in situations that are consistent with its intended use and scope; otherwise it may lead to confusing interpretations of data.

Our experience from complex case studies on coastal ecosystems and estuaries drove us to recognize that, despite their potential utility, most ecological indicators are more often than not relatively specific for a given kind of stress, or applicable to a particular type of community and/or scale of observation.

The idea of preparing a user-friendly guide for practitioners regarding the use of ecological indicators was therefore ripened. Our objective was, besides proving a world wide updated compilation of the potential indicators presently available to quantify the status of coastal and estuarine ecosystems, the building up of a decision tree to facilitate which indices should be chosen in any particular case involving benthic fauna. Finally, we illustrate the relative performances of various indicators in portraying visible qualitative differences among a suite of marine and estuarine ecosystems.

This book is therefore intended to serve a number of audiences, acting as a useful source for managers, policy makers, researchers, students, and the informed public in general, which will allow the transfer of results from several research projects to a wider number of potential end users.

The Authors, 2009

1   Introduction

1.1   What are indicators and what is their utility?

Ecological indicators are commonly used to provide synoptic information about the state of ecosystems. They mostly address the ecosystem's structure and/or functioning accounting for a certain aspect or component, for instance nutrient concentration, water flow, macroinvertebrate and/or vertebrate diversity, plant diversity and productivity, erosion symptoms and, occasionally, ecological integrity at a system level.

Indicators are quantitative representations of either the forces that steer an ecosystem, of responses to forcing functions or of previous, current, or future states of an ecosystem. When used effectively, indicators are expected to reveal conditions and trends that will help in developmental planning and decision-making processes.

The main attribute of an ecological indicator is the combination of numerous environmental factors in a single value, which may be useful in terms of management and the development of ecological concepts, compliant with the general public's understanding. Moreover, ecological indicators may help in establishing a useful connection between empirical research and modelling, since some of them are of use as orientors (also referred to in literature as goal functions) in ecological models (Jørgensen & Bendoricchio, 2001).

Such application proceeds from the fact that conventional models of aquatic ecosystems are not effective in predicting the occurrence of qualitative changes in ecosystems, for example shifts in species composition. This is due to the fact that measurements typically carried out, such as those of biomass and production, are not able to capture such modifications (Nielsen, 1995). Nevertheless, it seems possible to incorporate these types of changes in structurally dynamic models (Nielsen, 1992, 1994, 1995; Jørgensen et al., 2002) This allows for the improvement of existing models, not only in terms of increasing their predictive capability, but also by approaching a better understanding of an ecosystem's behaviour, and, consequently, a better environmental management.

In structurally dynamic models, the simulated behaviour and development of the ecosystem (Nielsen, 1995; Straškraba, 1983) is guided through an optimization process by changing the model parameters in accordance with a given ecological orientor (goal function). In other words, this permits introducing within models parameters that vary as a function of changing forcing functions and conditions of the state variables and optimizing model outputs in a step-by-step approach. In this case, orientors are assumed capable of capturing the macroscopic property of a given ecosystem by expressing emergent characteristics arising from self-organisation processes.

In general, the application of ecological indicators is not exempt of criticisms, the first of which is that aggregation results in an over-simplification of the eco-system under observation. Moreover, problems arise from the fact that indica-tors account not only for numerous specific system characteristics, but also other kinds of factors, such as physical, biological, ecological, socio-economic, etc. Therefore, indicators should forcibly be handled following the right criteria and in situations that are consistent with its intended use and scope; otherwise, they may lead to confusing interpretations of data.

1.2   What are the characteristics of a good indicator?

The level of quality of any given indicator will always remain a matter of perspec-tive. For instance, from a relatively holistic viewpoint, O'Connor & Dewling (1986) proposed, just over two decades ago, five criteria to define a suitable index for the assessment of ecosystem degradation, which in our opinion can still be considered up-to-date. An index should be: (1) relevant; (2) simple and easily understood in layman's terms; (3) scientifically justifiable; (4) quantitative and (5) cost effective. That same year, Hellawell (1986) detailed the following characteristics as being, from a toxicological perspective, ideal for an indicator species: (1) easy to identify and to sample; (2) have universal distribution; (3) be a resource of economic impor-tance; (4) easy to cultivate and maintain in laboratory conditions and (5) exhibit bio-accumulative ability and low genetic variability. Such features are obviously strictly related to the concept of bio-accumulator.

For the field ecologist, a good ecological indicator would be defined in terms of the following characteristics (Salas, 2002): (1) ease in handling; (2) sensitiv-ity to small variations of environmental stress; (3) independence of reference states; (4) applicability in extensive geographical areas and in the greatest pos-sible number of communities or ecological environments and (5) relevance to policy and management needs.

UNESCO (2003) also listed the characteristics that environmental indica-tors should present as: (1) have an agreed scientifically sound meaning; (2) be representative of an important environmental aspect for the society; (3) provide valuable information with a readily understandable meaning; (4) be meaningful to external audiences; (5) help in focusing on necessary information to answer important questions and (6) assist the decision-making process by being efficient and cost effective in terms of use.

Dale & Beyeler (2001) contributed by considering the following as the most-suit-able qualities of a good ecological indicator: (1) be easily measured; (2) be sensitive to stresses on the system; (3) respond to stress in a predictable manner; (4) predict changes that can be averted by management actions; (5) be anticipatory; (6) be inte-grative; (7) have a known response to natural disturbances, anthropogenic stresses and changes over time and (8) have low variability in its response.

Despite the evident convergence of ideas between different authors, there is no unanimous opinion on classifying an indicator as 'good'. Moreover, it is obviously not easy to fulfil all these requirements. Despite the panoply of bio-indicators and ecological indicators that can be found in the literature, they are more often than not specific to a given kind of stress, or applicable to a particular type of community and/or scale of observation, and rarely has their validity been proven.

1.3   Book structure

This work essentially addresses three tasks spread across five chapters: (1) the revision of the potential indices presently available to quantify the status of coastal and transitional ecosystems, namely under the scope of the European Water Framework Directive (WFD, 2000/60/CE); (2) the building up of a decision tree to facilitate which indices should be chosen in any particular case involving benthic fauna and (3) the evaluation of the performance of various indices in portraying visible qualitative differences among a suite of marine and estuarine ecosystems.

Following the introduction, Chapter 2 examines ecological indicators and their characteristics. It includes brief references to terrestrial and freshwater ecological indicators and a comprehensive review of those indicators applied in assessing coastal and marine environments. It contemplates six groups of indices: (1) indicators based on indicator species; (2) indicators based on ecological strategies; (3) indicators based on diversity; (4) indicators based on species biomass and abundance; (5) multimetric indices and (6) indices thermodynamically oriented or based on network analysis. Chapter 3 provides a decision tree for selecting ecological indicators as a function of benthic fauna data type and availability. Chapter 4 shows how this decision tree was applied to different case studies. Finally, Chapter 5, with the additional presentation of two case studies, discusses how to combine indicators when characterising a systems' ecological status, in the scope of the WFD implementation.

2 Review of ecological indicators and their characteristics

There are numerous ecological indicators designed to measure the health status of terrestrial as well as aquatic (freshwater or marine) environments. To carry out a comprehensive review of all of them is certainly out of our range, but the development of a guide concerning the right use of such indices, taking as examples some of those environments, will constitute a good tool for further studies regarding its application in environmental management. In the present work, although the most commonly applied terrestrial and freshwater ecological indicators are concisely approached, we chose to concentrate on those applied in the environmental quality assessment of estuaries and marine ecosystems.

Most of the ecological indicators used and/or tested in evaluating the health status of marine and transitional water ecosystems can be found in the literature and are the result of just a few distinct theoretical approaches. A first group of indicators focuses on the presence/absence of a given indicator species, while others take the different ecological strategies adopted by organisms into account, as well as its diversity, or the energy variation in the ecosystem resulting from changes in the individuals' biomass. A second group of indicators is thermodynamically oriented or based on network analysis and aims at gathering the information on an ecosystem from a more holistic perspective. Finally, a third group attempts to include information regarding a given environment in one single value, constituting the so-called multimedia indices.

Indicators based on diversity, as well as those that are thermodynamically oriented, can be used in all types of systems. Indices based on indicator species and ecological strategies, as well as multimetric indices, are, on the other hand, more often specifically designed as a function of the environment to be evaluated, despite sharing the same conceptual bases.

2.1 Brief reference to terrestrial ecological indicators

The search for biological indicators of disturbance in terrestrial environments has been undertaken in different directions (Blair, 1996; Mason, 1996; McGeoch, 1998). For instance, terrestrial invertebrates are good indicators because they are ubiquitous, diverse, easy to sample and ecologically important (Andersen, 1997). They play diverse roles in natural environments as decomposers, predators, parasites, herbivores and pollinators, and respond to various perturbations (Price, 1988). Additionally, certain taxa such as beetles, butterflies, spiders and ants respond to effects of human or natural disturbance.

Therefore, some groups such as Formicidae (ants) and Carabidae (beetles) have already been thoroughly studied as disturbance indicators. Carabid beetles, for instance, respond to agricultural practices, fires and clear cutting (Refseth, 1980; Holliday, 1991; Niemela *et al.*, 1993). Within this group, different trophic groups show different sensitivity levels to agricultural management. For instance, it has been observed that carnivore and phytophage taxa richness tends to decrease rapidly when disturbances lead to landscape simplification, while that for polyphagous taxa might even increase because of their opportunistic feeding habitats and a higher tolerance to disturbance factors (Purtauf *et al.*, 2005).

Ants have been used in Australia and in the United States of America in monitoring programmes associated with mining, fires, grazing and logging, and according to several authors (e.g. Majer *et al.*, 1984; Neumann, 1992; Andersen, 1997; Nash *et al.*, 1998) taxa richness is higher in some types of disturbed sites. Spiders, in turn, are affected by vegetation architecture and prey availability (McIver *et al.*, 1992). However, some spiders such as the wolf spider are better adapted to disturbance because since they carry their egg sacs with them, they are able to colonize disturbed areas (Uetz, 1976). Lepidoptera include taxa with diverse trophic roles (Hammond & Miller, 1998) and according to several studies (e.g. Holl, 1996; Spitzer *et al.*, 1997) the occurrence of fewer taxa at disturbed sites is less likely.

Kimberling *et al.* (2001) designed a biological integrity index based on terrestrial invertebrates in the shrub-steppe of eastern Washington (United States of America). This index accounts for the following eight metrics: total number of families; number of Diptera families; relative abundance of detritivores; taxa richness of Acaria; predators; detritivores; ground-dwellers and polyphagous carabid beetles.

Regarding terrestrial vertebrates, birds have been found to be useful biological indicators because they are ecologically versatile and respond to secondary changes resulting from primary causes, and they can be monitored relatively inexpensively (Koskimies, 1989). In addition, because of their degree of mobility, birds react rapidly to changes in their habitat (Morrison, 1986; Fuller *et al.*, 1995; Louette *et al.*, 1995). According to Browder *et al.* (2002), bird taxa are appropriate indicators in monitoring changes for several reasons: (1) individual bird species are associated with particular habitats; (2) birds can be found across a broad gradient of anthropogenic disturbance, from pristine wilderness to metropolitan areas; (3) most bird species live only a few years, so changes in species composition and abundance will manifest themselves relatively quickly after a disturbance; (4) groups of bird species can be used to develop associations with habitats that are predictive of the relative level of anthropogenic disturbance and (5) birds are important to a large segment of the public (Szaro, 1986; Canterbury *et al.*, 2000), who will accordingly relate more to concerns about changes in bird communities than those that involve other taxa such as plants or invertebrates.

Browder *et al.* (2002) developed a measure of grassland integrity using the presence and abundance of disturbance-intolerant and disturbance-tolerant bird species. This index provides a method of monitoring grassland integrity based on the tolerance of grassland birds to anthropogenic disturbance, particularly culti-vation. On the other hand, Reynaud & Thiolouse (2000) used co-inertia analysis to identify birds as biological markers along an urban–rural gradient.

In plant communities, vegetation cover is generally used to measure bio-logical diversity and to detect anthropogenic disturbances such as the change from high-diversity prairies and late successional forests dominated by peren-nial native species to relatively homogeneous agricultural fields dominated by annual crops and weed species (Delong & Brusven, 1998). Understorey weeds have been used as effective indicators of deciduous-forest regeneration in southern Canada (McLachlan & Bazely, 2001), of long-term continuity of the boreal forest in Sweden (Ohlson *et al.*, 1997), of military traffic in longleaf pine forests in Georgia (Dale *et al.*, 2002) and of riparian forest disturbance in the southern United States of America (Bratton *et al.*, 1994).

Diversity measures, such as total species richness, are often used as indica-tors of forest changes but require a full characterization of the forest (Moffatt & McLachlan, 2004). Although the use of individual plant species as indicators may eliminate the need for a full description of forests, they may only yield site-specific information and reveal little about the mechanisms underlying for-est change. The use of guilds, groups of species that are functionally related and have similar resource requirements, represents an intermediate solution for describing disturbance impacts (Hobbs, 1997). Thus, life history and life form have been used to monitor forest disturbances (McIntyre *et al.*, 1995; Dale *et al.*, 2002), and origin and habitat preference have been related to forest species loss and compositional change associated with urban land use (Drayton & Primack, 1996). Flowering phenology and seed dispersal have been related to the decline of species associated with human usage (McLachlan & Bazely, 2001).

Moffatt & McLachlan (2004) showed that herbaceous species, both individu-ally and grouped according to functional types or guilds, are effective indicators of environmental change and disturbance associated with land use. They also identified two categories of species' responses to urban land use: urban exploiters, restricted to or dominant in disturbed urban forests; and urban avoiders, excluded from disturbed urban forests. A third set of plant species appeared in association with both urban and suburban sites, in contrast to a fourth group that was more frequent in rural and reference sites. In addition, they observed that most of the indicators of disturbance and opportunistic species were exotic, while nearly all vulnerable species were native, as were all species identified as effective indicators of forests in a state of integrity.

On the other hand, life-history traits also lie beneath understorey responses to land use. For instance, woody species tend to be more resistant to disturbance, perhaps due to a relatively longer life span and a greater structural durability (Robinson *et al.*, 1994). On the other hand, annuals tend to respond positively to disturbance, in part because of their often rapid rates in terms of biomass and abundant seed production (Bazzaz, 1986).

Seed dispersal is also subjacent to understorey responses to land use. Dis-turbance indicator species tend to be endozoochores that produce berries. Other studies of degraded forests have found that myrmecochores (Dzwonko & Loster, 1992), barcoheres (Matlack, 1994) and ephemerals (McLachlan & Bazely, 2001) are vulnerable to habitat fragmentation and physical disturbances.

Wind-dispersed seeds are also likely to exhibit a higher mortality rate in greatly fragmented urban environments because their dispersal patterns are largely non-selective (Van der Pijl, 1972). Dispersal-restricted species (those that are gravity, explosion or ant-dispersed) often travel only a few centimetres per year and are usually unable to traverse the large gaps that separate urban patches (Dzwonko & Loster, 1992).

DeKeyser *et al.* (2003) developed an Index of Plant Community Integrity (IPCI) to quantitatively assess the seasonal wetland communities. They delineated plant data into the same metrics as the rest of the data set (e.g. species richness, percentage of introduced and annual plants) and analysed these metrics using principal components and cluster analyses, which allowed the defining of five quality classes: Very Good, Good, Fair, Poor and Very Poor.

In other cases, some measurements, like the Fluctuating Asymmetry (FA) in some vegetal species such as *Lythrum salicaria*, have been used to detect pollution by heavy metals (Mal *et al.*, 2002). FA measures the random deviation from perfect bilateral or radially symmetrical morphological traits in a group of organisms (Wilsey *et al.*, 1998; Palmer & Strobeck, 1986). In a bilaterally symmetrical trait, the left side is identical to the right. Deviations from perfect bilateral symmetry can be in the form of directional asymmetry, antisymmetry or FA (Palmer, 1994; Leary & Allendorf, 1989). Of the three kinds of asymmetry mentioned above, only FA is thought to be caused by developmental noise or imperfect developmental stability (Palmer, 1994; Palmer & Strobeck, 1986) and several studies have shown that it is related to abiotic stress, noise, nutrition and pollutants (Möller & Swaddle, 1997). In fact, the FA in three species (*Robinia pseudocaia*, *Sorbus aucuparia* and *Matricaria perforata*) increased with the amplified levels of radiation in Chernobyl (Möller, 1998). Zvereva *et al.* (1997) and Kryazheva *et al.* (1996) have also reported an increment in FA of leaves caused by metal and chemical pollution in the air.

Functional indices have also been used in terrestrial ecosystems. A food web approach to disturbance and ecosystem stress was applied by Moore & de Ruiter (1997a, 1997b) and food webs and productivity, combined with nutrient cycling, have been used to assess stability and disturbance of soil ecosystems and agrosystems in a number of cases (e.g. Moore & de Ruiter, 1991; Moore *et al.*, 1993; de Ruiter *et al.*, 1994, 1995).

2.2   Brief reference to freshwater ecological indicators

2.2.1   Macroinvertebrates

The use of freshwater macroinvertebrates in biomonitoring programmes is well established (Rosenberg & Resh, 1993) with a multitude of different approaches based on biotic indices, indicator species, species diversity, toxicology, community composition or ecosystem function (Resh & McElravy, 1993).

With regard to freshwater aquatic environments, the use of biotic indices is the most common approach in assessing the quality of a river or lake. The biotic-approach assessment, as defined by Tolkamp (1985), is one which combines diversity on the basis of certain macroinvertebrate taxonomic groups with the pollution indicators of individual species or higher taxa or groups into a single index or score (Calow & Petts, 1992). Numerous biotic indices and score systems have been developed. According to Calow & Petts (1992), the most widely used systems till the 80s are illustrated in Figure 1.

The *Trent Biotic Index (TBI)* (Woodiwiss, 1964) was originally devised for use in the Trent River Authority area in England, but has since been adapted for use in many other countries and appears to form the basis for most modern biotic indices and scores (Persoone & De Pauw, 1979). The *Chandler's Score System* (Cook, 1976) was developed for upland rivers in Scotland and is theoretically an improvement over the TBI, because it includes an abundance factor and incorporates a more detailed list of macroinvertebrates (Calow & Petts, 1992).

The *Belgian Biotic Index (BBI)* (De Paw & Vanhooren, 1983) is derived from the French *Biotic Index (IB)* (Chutter, 1972), which in turn is a modification of the TBI. The BBI is based on the total number of systematic units and number of units in different faunal groups, and the IB determines a community score by weighting the relative abundance of each taxa in terms of its tolerance to pollution.

There is no doubt that the most well-known index in Europe is the *Biological Monitoring Working Party (BMWP)*. The BMWP was set up in 1976 by the UK Department of Environment to develop a standardised system for assessing the biological quality of rivers in Great Britain. They developed a standardized score system that was a simplification of Chandler's Score (where all organisms were identified familywise for uniformity; families with similar pollution tolerances were grouped together and the abundance factor was eliminated because it was time consuming and had only a small effect on score values (Calow & Petts, 1992)). To apply the BMWP, it is necessary to list the families present in the sample, ascribe the score for each family and then add the scores together to arrive at a site score (for more details see Armitage *et al.*, 1983 and Hellawell, 1986).

The BMWP suffered a number of subsequent modifications accounting for the taxa found in the geographical areas where it was applied. For instance, in the Iberian Peninsula, the index was modified by Alba-Tercedor & Sánchez-Ortega (1988) and was named *BMWP'*. After several revisions (Alba-Tercedor, 1996; Alba-Tercedor & Pujante, 2000), Alba-Tercedor *et al.* (2002) named it as *Iberian Biomonitoring Working Party (IBMWP)*. The index is computed adding the punctuations attributed to the different taxa found in macroinvertebrate samples, which are cited in a list developed for this purpose. The punctuation assigned to a given taxa is proportional to its higher or lower sensitivity to organic pollution and to the level of oxygen deficit usually resulting from that type of pollution in most rivers, with the exception of the most torrential ones, where water agitation determines higher oxygenation.

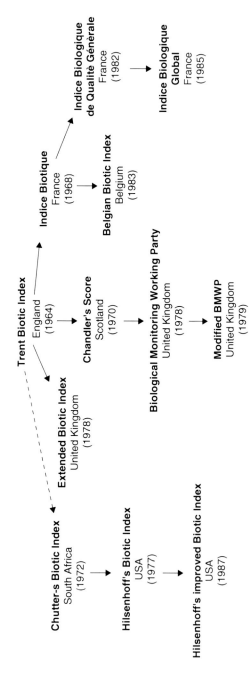

Figure 1: Development of the most widely used biotic indices and score systems (till the 80s) (after Calow & Petts, 1982).

Other biotic indices were developed across a wide range of geographical regions, for example the *Florida Index (FI)* (Ross & Jones, 1979) that is a weighted sum of intolerant taxa (insects and crustaceans) found at a site, the *Hilsenhoff's* (1987, 1988) *Wisconsin Biotic Index* (BI) and *Family Biotic Index (FBI)*; the *Chessman's* (1995, 1997) *Australian Stream Invertebrate Grade Number – Average Level* (SIGNAL-95 and SIGNAL-HU97). *New Zealand's Macroinvertebrate Community Index (MCI)*, its quantitative variant *(QMCI)* (Stark, 1985) or the *Semiquantitative Macroinvertebrate Community Index (SQMCI)* (Stark, 1998) to assess organic enrichment in stony riffles in streams and rivers. The above-mentioned biotic indices are only a small portion of the available panoply.

Other biological integrity indices account for the invertebrate populations. An example is the *Invertebrate Community Index* (EPA, 1987), calculated as the sum of 10 individual measures (total number of taxa, total number of Trichoptera taxa, total number of Diptera taxa, percentage of Ephemeroptera, percentage of Trichophera, percentage of the tribe Tanytarsini of the Chironomidae, percentage of other Dipterans and non-insects, percentage of tolerant organisms and number of Ephemeroptera, Plecoptera and Trichoptera taxa) that are scored individually. The *Mean Biometric Score* (Shackleford, 1988) is also a combination of community diversity, indicator organisms and approaches of functional groups, and the *Biological Condition Score* (Plafkin *et al.*, 1989) is calculated through eight metrics that reflect the groups' tolerance, community structure and community function.

More recently, in the scope of the WFD implementation, several biotic indices are being used, the IBMWP, the *Invertebrate Portuguese Index (IPtI)* (EC, 2007a), the *STAR Intercalibration Common Metric Index (STAR_ICMi)* (Buffagni *et al.*, 2007), the *Indice Biologique Global Normalisé (IBGN)* (AFNOR, 1992) or the *ICMStar* (Buffagni *et al.*, 2005, 2006, 2007)

Alternative approaches to biotic indices have been developed by other authors. Namely, Zelinka & Marvan (1961) have proposed a *Saprobic Index (S)*, which is based on the number and abundance of the taxa included in the saprobe list or the ISO Score (ISO, 1984), calculated as the sum of the tolerance scores for the taxonomic families present.

With regard to indices based on ecological strategies used in freshwater systems, we may refer: (1) the *ratio of shredders in relation to the total number of individuals* (Plafkin *et al.*, 1989) based on the assumption that shredder organisms and their microbial food base are sensitive to toxicants and the modification of the riparian zone; (2) the *ratio of scrappers to collector–filterers*, which assumes that collector–filterer dominance may reflect organic enrichment; (3) the *ratio of trophic specialists in relation to generalists* (Maine Department of Environmental Protection, 1987), which considers trophic generalists to be more pollution-tolerant, thus becoming numerically dominant in response to environmental stress and (4) the *ratio of EPT abundance in relation to Chironomidae abundance*, which accounts for the fact that Chironomidae are perceived to be pollution-tolerant as compared to the pollution-sensitive Ephemeroptera, Plecoptera and Trichopera. Compared to a non-stressed habitat, a stressed one will show an unbalanced composition regarding these groups (Resh & Jackson, 1993).

Apart from the indices described, diversity indices are also used. Diversity indices are mathematical expressions which use three components of community structure, namely richness (number of species present), evenness (uniformity in the distribution of individuals among the species) and abundance (total number of organisms present) to describe the response of a community to the quality of its environment (Calow & Petts, 1992). So far the most widely used diversity index is the *Shannon–Wiener Index (H')* (Shannon & Weaver, 1963). Nevertheless, the *Berger–Parker Index (d)* (Berger & Parker, 1970; May, 1975), *MacArthur's Diversity Index (D)* (MacArthur, 1972), the *Margalef Index (d)* (Margalef, 1969), the *Pielou Index (J)* (Pielou, 1969) and the *Simpson Index (D)* (Simpson, 1949) are also commonly used (for more details see Chapter 2, Section 2.3.3.).

Among the most widely indices used to measure the biological diversity in rivers, one of the most interesting is the Fractal Dimension of Biocenosis (*D*), proposed by Margalef (1991), and originally developed by Docampo & Bikuña (1991) as a biological index to be applied in assessing river communities. Its present formulation expresses the speed in identifying the benthic invertebrate species or other taxocenosis when the size of the biological sample increases (number of collected individuals or number of analysed individuals), according to the following equation *LogS/LogN*, where: *S* is the richness in species, or alternatively, the taxonomic richness; and *N* is the number of individuals. In non-polluted rivers, it holds an average value of 0.385 and decreases strongly in rivers impacted by human influence.

2.2.2   Phytobenthos

There is often the need to determine the phyto-physiological status of fluvial stretches in terms of quality diagnosis accounting for the different behaviours of water masses as a function of their response to the increment of primary producers (microphytes and macrophytes). This assessment is carried out according to which of the two phyto-physiological types considered in river ecosystems they belong to. These two types, photosystem I and photosystem II, are established based on photo-pigment concentration that is used to diagnose the phyto-physiological status of the river system. Photosystem I is characterised by a high value of the *a/b* index on *Cladophora*, implying a high dominance of algae in the fluvial ecosystem, an oversaturation state of chlorophyll *a* (vegetal biomass), and, concurrently, eutrophication or even hypereutrophication. Photosystem II, on the other hand, is characterised by a low value for the *a/b* index, which implies natural algae metabolism conditions, with a balanced ratio between production and assimilation in the system. *Margalef's Pigment Index* (1989), *D430/D665*, which measures the relation between the concentration of all the pigments (carotenes, xanthophylls, as well as *a*, *b*, *c* and *d* chlorophylls) as well as the concentration of chlorophylls alone, can be used to distinguish between the two photosystems. Lower values are got when chlorophyll *a* is predominant (System I), and increasing when the other pigments are well represented (System II) or when chlorophyll *a* degrades, increasing the degradation products (among which pheophytine can be found (Margalef, 1983)).

The index can therefore show abnormally high values in polluted rivers (Margalef, 1983). However, the *Chlorophyte Index, IC* (which expresses the cologarithm of the ratio *a/b*) provides a clearer distinction between both types of photosystems as compared to the D430/D665 index. Negative values of IC indicate that the stretch bears photosystem I characteristics, and positive values indicate photosystem II.

Ecological studies have long since made it clear that algae, especially benthic algae that possess limited mobility, constitute one of the best indicators of the conservation status and biological quality of aquatic systems (Lowe & Pan, 1996). Diatoms are important contributors of the primary production in aquatic ecosystems (Wetzel, 1983). They can serve as good indicators of the ecological status of surface waters. Biotic measures have advantages against chemical analyses. Most importantly, they integrate environmental effects (Cox, 1991) reflecting the typical conditions instead of momentary values that can be measured precisely with chemical methods. Diatom-based impact analyses have a long history (Kolkwitz & Marson, 1908; Butcher, 1947). A number of methods were developed for the use of diatoms as bioindicators of changing environment, especially in rivers. Several European countries (Poland, Germany, France, Austria, Switzerland and United Kingdom) currently maintain a control net based on the use of diatoms. The *Diatom Biological Index (IBD)* (Coste in Cemagref, 1982), in particular, was developed by the French water agencies with the aim of implementing throughout the country a method first developed in the Siena basin, and in the Rodane–Mediterranean–Corsica and Artois–Picardie water agencies. The IBD is based on 209 taxa and seven water-quality classes defined from 14 common physical and chemical parameters, its values range from 1 (very bad quality) to 20 (very good quality) (for more details regarding the calculations see AFNOR, 2000 and Prygiel & Coste, 2000).

The *Trophic Diatom Index (TDI)* (Kelly & Whitton, 1995; revised by Kelly *et al.*, 2001) has been developed, in response to the National Rivers Authority's (England & Wales) needs under the terms of the Urban Wastewater Treatment Directive of the European Community. The index is based on a suite of 86 taxa selected both for their indicator value and ease of identification. When tested on a dataset from 70 sites free of significant organic pollution, this index was more highly correlated with aqueous P concentrations than previous diatom indices. However, where there was heavy organic pollution, it was difficult to separate the effects of eutrophication from other effects. Its values varied from 1 (very low organic concentration) to 5 (very high organic concentration). Another index sensitive to nutrients is the Austrian *Trophic Index* developed by Rott *et al.* (1999).

Other two diatom water-quality indices, considered to be global pollution indicators (Prygiel & Coste, 1999), are the *Specific Polluosensitivity Index (IPS)* (Coste in Cemagref, 1982) and the *Index of European Economic Community (CEE)* (Descy & Coste, 1990, 1991). The IPS is based on the Zelinka & Marvan (1961) formula. Conversely, the CEE index is based on a twofold quality grid. Different numbers of taxa are taken into account by each index: CEE uses 208 taxa, and the IPS uses all known taxa. Both evaluate the water quality in a range between 1 and 20, 1 indicating a very bad and 20 a very good water quality. The *Intercalibration Common Metric (ICM)* (EC, 2007b) was derived during the

WFD Intercalibration exercise and is essentially composed by the IPS (Coste in Cemagref 1982) and Trophic Index (TRIX) (Rott *et al.*, 1999).

2.2.3 Fish

Fish communities are also used in assessing the quality of hydrographical basins. They are subject to a wider variety of impacts than benthic macro-invertebrates, such as species extinction as a consequence of mechanical and physicochemical pollution of the waters, population movements and the favouring of allochthon fish species in detriment of autochthon species due to water nutrient enrichment and also the reduction in the circulating volume of water, and the channelling of fluvial stretches. Particularly frequent is the risk of competitive exclusion of the autochthon species at a local or metapopulational level deriving from the presence of alochthon species (Borja *et al.*, 2003a). Among the most-cited indices based on fish communities, it is worth referring to the Index of *Biotic Integrity (IBI)* (Karr, 1981). The IBI principle is based on the fact that fish communities respond to human alterations of aquatic ecosystems in a predictable and quantifiable manner. An IBI is a tool to quantify human pressures by analysing alterations of the structure of fish communities. The original IBI (Karr, 1981) uses several components of fish communities, for example taxonomic composition, trophic levels, abundance and fish health. Each component is quantified by metrics (e.g. proportion of intolerant species).

Currently, different fish-based methods are used in Europe, while most countries have not yet included fish in their routine monitoring programmes. Thus, the successful implementation of the WFD depends on the provision of reliable and standardized assessment tools. This was the motivation for the EC-funded FAME project to develop, evaluate and implement a fish-based assessment method for the ecological status of European rivers to guarantee coherent and standardized monitoring throughout Europe. The *European Fish Index (EFI)* (FAME, 2004), founded on the concept of the IBI (Karr, 1981), is based on a predictive model that derives reference conditions for individual sites and quantifies the deviation between predicted and observed conditions of the fish fauna. The ecological status is expressed as an index ranging from 1 (High ecological status) to 0 (Bad ecological status). EFI uses data from single-pass electric fishing catches to calculate the assessment metrics. The EFI employs 10 metrics belonging to the following ecological functional groups: trophic structure (density of insectivorous species, density of omnivorous species); reproduction guilds (density of phytophilic species, relative abundance of lithophilic species); physical habitat (number of benthic species, number of rheophilic species); migratory behaviour (number of species migrating over long distances) and capacity to tolerate disturbance in general (relative number of intolerant species, relative number of tolerant species). The final EFI is obtained by summing the ten metrics, and then by rescaling the score from 0 to 1. The final step is to assign index scores to ecological status classes.

2.3   Review of the ecological indicators used in assessing coastal and marine environments

Following the promulgation of the European Water Framework Directive (WFD, 2000/60/CE), the need for stable and comparable criteria in environmental quality assessment of aquatic ecosystems, including coastal zones and estuaries, reactivated the use and search of ecological indicators of pollution.

In this review, we consider the indices most used to assess pollution effects in transitional and coastal areas. The algorithms of the different indices are provided in full detail, and their application in different scenarios, with regard to the necessary requirements as a function of data quality and availability, will be further approached through a binary key (see Chapter 3).

2.3.1   Indicators based on indicator species

Amongst those that are usually denominated indicator species, we are able to determine two distinct cases, whether they are considered as indicators in the most common sense, or as bioaccumulator species (the latter more appropriate in toxicological studies), which may sometimes lead to confusion. In the first case, we are referring to those species whose appearance and dominance is associated with an environmental deterioration, because they are favoured for that feature, or because they are more tolerant to that type of pollution than other less-resistant species. In this sense, the possibility of assigning a certain grade of pollution to an area in terms of the species present has been pointed out by a number of researchers (Blegvad, 1932 & Filice, 1954 in Planas & Mora, 1987; Glémarec & Hily, 1981), mainly with regard to organic pollution studies. In fact, different authors have focused on the presence/absence of such species to formulate biological indices. For instance, the Bellan Index (based on polychaetes), or the Bellan–Santini Index (based on amphipods), attempts to characterise environmental conditions by analysing the dominance of species indicating some type of pollution in relation to the species considered to indicate an optimal environmental situation (Bellan, 1980; Bellan–Santini, 1980).

Nevertheless, many authors claim that the use of such indicators is not advisable because, more often than not, the species being examined may occur naturally in relatively high densities. In fact, there is no reliable methodology to know at which level any one of those indicator species can be well represented in a community that is not really affected by any kind of pollution, which leads to a significant exercise of subjectivity (Warwick, 1993). Despite these criticisms, even recently, the *AMBI* Index (Borja *et al.*, 2000), based on the Glémarec & Hily (1981) species classification regarding their response to pollution, the BENTIX Index proposed by Simboura & Zenetos (2002), the Norwegian Indicator Species Index (ISI) (Rygg, 2002) as well as the Benthic Quality Index (BQI) (Rosenberg *et al.*, 2004), all applying the very same principles, have had to update such pollution-detecting tools. Moreover, Roberts *et al.* (1998) also proposed an index based on macrofaunal species that accounts for the ratio of each species' abundance in control versus samples proceeding from stressed areas. This proposal is, however, semi-quantitative, as well as site- and pollution-type specific. In the same way, the Benthic Response Index (BRI) (Smith *et al.*, 2001) is based on the type of species present in a sample (related to pollution tolerance), but its applicability is complex seeing as it is calculated using a two-step process in which ordination analysis is employed to quantify a pollution gradient within a calibration data set.

The *AMBI* Index, for instance, which accounts for the presence of species indicating a given type of pollution, as well as species indicating a non-polluted situation, has been considered very useful in terms of implementing the WFD in coastal ecosystems and estuaries. In fact, although this index is very much based on the paradigm of Pearson & Rosenberg (1978), which emphasizes the influence of organic matter enrichment on benthic communities, it has established its usefulness in the assessment of other anthropogenic impacts, such as the physical disturbance of a habitat, heavy-metal inputs, etc. It has, in fact, been successfully applied in the Atlantic (North Sea, Bay of Biscay and South of Spain) and in the Spanish and Greek European Mediterranean coasts (Borja *et al.*, 2000; 2003b; 2003c; Casselli *et al.*, 2003; Forni & Occhipinti Ambrogi, 2003; Nicholson & Hui, 2003; Bonne *et al.*, 2003; Muxika *et al.*, 2003; Gorostiaga *et al.*, 2004; Salas *et al.*, 2004).

Marine benthic macrophytes, in their turn, respond directly to the abiotic and biotic aquatic environments, and thus represent sensitive bioindicators regarding their changes (Orfanidis *et al.*, 2003). On the other hand, a series of algae genera are universally considered to appear when pollution occurs, such as the green algae *Ulva, Cladophora* and *Chaetomorpha* and the red algae *Gracilaria, Porphyra* and *Corallina*. Additionally, species with high structural complexity, such as the Phaeophyta belonging to *Fucus* and *Laminaria* genera, are seen worldwide as the most sensitive to any kind of pollution, even if *Fucus* species may cope with moderate pollution (Niell & Pazó, 1978). Finally, marine Spermatophytae are considered indicator species of good water quality.

In the Mediterranean Sea, for instance, the presence of *Cystoseira* and *Sargassum* (Phaeophyta) or *Posidonia oceanica* meadows indicate the water's good quality. Thus, monitoring population density and distribution of such species allows detecting and evaluating the impact of all kinds of activities (Pérez-Ruzafa, 2003). *P. oceanica* is possibly the most-used indicator of water quality in the Mediterranean due to its sensitivity to disturbances, its wide distribution along the Mediterranean coast and the good knowledge about the plant and its ecosystem-specific response to a particular impact (e.g. Ruiz *et al.*, 2001; Pergent-Martini *et al.*, 2005; Romero *et al.*, 2005). Furthermore, this species is capable of giving information about the present and past level of trace metals in the environment (Pergent-Martini, 1998).

Pergent-Martini *et al.* (2005) identified the descriptors of *P. oceanica*, constituting the first step in opening the way to the use of this species to assess the ecological status of Mediterranean coastal zones (Table 1). POMI, the *P. oceanica* multivariate index, was developed on these grounds and based on those physiological, morphological and structural descriptors combined into a variable using a PCA (see Romero *et al.*, 2005).

Table 1: Recompilation of the main descriptors of *Posidonia oceanica* (Pergent-Martini *et al.*, 2005).

Descriptor	Measured parameters	Information
Upper depth limit	Depth, localization, density, bottom cover, characterization of the substrate	Human impact, hydrodynamism, sedimentary dynamics
Density	Number of shoot on a surface >1600 cm^2	Dynamics of the meadow, Human impact
Epiphytic coverage	Biomass, diversity	Nutrient concentration, flora and fauna biodiversity
Bottom cover	% of meadow on a given surface (1–25 m^2)	Dynamics of the meadow, Human impact
Leaf biometry	Type, number, size of leaves, leaf surface, Coefficient A, biomass, epiphytic coverage, presence of necrosis	Health state of the meadow, Human impact, hydrodynamism, herbivore pressure
Lower depth limit	Depth, localization, type, density, bottom coverage, leaf biometry, granulometry, content in organic matter	Water transparency, human impact, hydrodynamism, dynamics of the meadow (regression of colonization)
Population associated to the meadow	Fauna, flora, diversity	Biodiversity, meadow–population interactions
Structure of the matte	Intermattes, 'cliff of dead matte', erosive structures, receding, silting up, biodiversity of the endofauna, homogeneity, resistance and compactness, % of plagiotropic rhizomes, width of the matte, physicochemical composition	Dynamics of the meadow, human impact, sedimentary dynamics, study of currents
Biochemical and chemical composition	Elementary composition (C, N, P) phenolic compounds, proteins, carbohydrates, stress enzymes	Dynamics of the meadow, Human impact, herbivore pressure
Datation measurement	Lepidochronology, plastochrone interval, paleo-flowering, primary production	Temporal evolution of the production, sedimentation speed, intensity of sexual reproduction, dynamics of the meadow, Human impact
Contamination	Metals (Hg, Cu, Cd, Pb, Zn)	Human impact

In the same sense, a Conservation Index (Moreno *et al.*, 2001), based on the named marine Spermatophyta, is used in the Mediterranean coasts. Along the same lines, Orfanidis *et al.* (2001) introduced a new Ecological Evaluation Index (EEI) to assess the ecological status of transitional and coastal waters in accordance with the WFD. This index is based on the marine benthic macrophyte classification in two ecological state groups (ESGs I,II), representing alternative ecological states (pristine and degradated). More recently and also in the scope of the WFD implementation, Scanlan *et al.* (2007) proposed the Opportunistic Macroalgae Assessment Tool.

2.3.1.1 Measures based on indicator species

2.3.1.1.1 Annelida pollution index (Bellan, 1980)

$$\text{API} = \sum \frac{\text{Dominance of pollution indicators}}{\text{Dominance of clean water indicators}}$$

Bellan (1980) considered *Platynereis dumerilli*, *Theosthema oerstedii*, *Cirratulus cirratus* and *Dodecaceria concharum* as pollution indicator species and *Syllis gracillis*, *Typosyllis prolifera*, *Typosyllis* spp and *Amphiglena mediterranea* as clear water indicators.

Index values above 1 show that the community is pollution disturbed. As organic pollution increases, the index values become higher allowing, in theory, to establish different pollution grades, although the author does not define them.

This index was, in principle, designed to be applied on rocky superficial substrates. Nevertheless, Ros *et al.* (1990) modified it to be equally applied to soft bottoms, considering other indicator species. In this case, the pollution indicator species are *Capitella capitata*, *Malococerus fuliginosus* and *Prionospio malmgren*, and the clear water indicator species is *Chone duneri*.

2.3.1.1.2 AZTI' Marine biotic index (*AMBI*) (Borja *et al.*, 2000)   The AZTI Marine Biotic Index relies on the distribution of individual abundances of the soft-bottom communities into five ecological groups (Grall & Glémarec, 1997):

Group I: Species very sensitive to organic enrichment and present under unpolluted conditions.

Group II: Species indifferent to enrichment, always present in low densities with non-significant variations with time.

Group III: Species tolerant to excess organic matter enrichment. These species may occur under normal conditions; however, their populations are stimulated by organic enrichment.

Group IV: Second-order opportunistic species, adapted to slight to pronounced unbalanced conditions.

Group V: First-order opportunistic species, adapted to pronounced unbalanced situations.

The species were distributed in those groups according to their sensitivity to an increasing stress gradient (enrichment of organic matter) (Hily, 1984; Glémarec, 1986). This index is based on the percentages of abundance of each ecological group of one site (Biotic Coefficient: BC), which is given by:

$$BC = \left\{ \frac{(0 \times \%G_I) + (1.5 \times \%G_{II}) + (3 \times \%G_{III}) + (4.5 \times \%G_{IV}) + (6 \times \%G_V)}{100} \right\}$$

The AZTI' Marine Biotic Index, also referred to as BC, varies continuously from 0 (unpolluted) to 7 (extremely polluted) (Table 2).

To implement this index, more than 3000 taxa have been classified, representing the most-important soft-bottom communities present in European estuarine and coastal systems. The AZTI' Marine Biotic Index can be applied using the AMBI© software (Borja et al., 2003b and www.azti.es, where the software is freely available).

It is possible to detect the impact of anthropogenic pressures in the environment with this index because it can be used to measure the evolution of the ecological status of a particular region. For example, Muxika et al. (2005) have tested it in different geographical sites such as the Basque Country coast-line, Spain, for where it was originally designed (Borja et al., 2000), the Mondego estuary, Portugal (Salas et al., 2004), three locations on the Brazilian coast and two on the Uruguayan coast (Muniz et al., 2005), and has been tested among different geographical sites (Muxika et al., 2005), allowing correct evaluations of the ecosystem's conditions. So, this index can constitute a sound tool for management due to its capacity to assess ecosystem health.

One drawback of *AMBI* is that mistakes can occur during the grouping of the species into different groups according to their response to pollution situations. Once it draws on the response of organisms to organic inputs in the ecosystem it does not detect the effects caused by other types of pollution, as for instance toxic pollution (Marín-Guirao et al., 2005). Moreover, this index presents some limitations when applied to semi-enclosed systems (Blanchet et al., 2008).

Table 2: Categories considered as a function of *AMBI* Index values.

Classification	*AMBI* value
Normal	0–1.2
Slightly polluted	1.2–3.2
Moderately polluted	3.2–5
Highly polluted	5–6
Very highly polluted	6–7

2.3.1.1.3   Benthic Quality Index (Rosenberg *et al.*, 2004)

$$\text{BQI} = \left(\sum_{i=1}^{n} \left(\frac{A_i}{\text{tot}A} \times Es50_{0.05i} \right) \right) \times^{10} \log(s+1)$$

Tolerant species are by definition predominantly found in disturbed environments. That means that they mainly occur at stations with low $ES50$, where ES is the diversity value measured by the Hulbert Index and s the mean number of species. In contrast, sensitive species usually occur in areas with minor disturbance or none at all, being then associated with high $ES50$ values. Taking into account the abundance frequency distribution of a particular species in relation to the $ES50$ values at the stations where it has been recorded, the most-tolerant individuals of a species are likely to be associated with the lowest $ES50$ values. The authors estimated that 5% of the population will be associated with this category, and defined this value as the species tolerance value: $ES50_{0.05}$.

The tolerance value of each species found at a given station is then multiplied by the average relative abundance (A) of that species (i), to weight the common species in relation to the rare ones. The sum is then multiplied by the $\log_{10}$ of the mean number of species (s) at that station, since higher species diversity is assumed to be related to better environmental quality. All information related to the number of species and their abundance at a given station is therefore used for this quality assessment. This index has only been applied in the Baltic Sea.

2.3.1.1.4   BRI (Smith *et al.*, 2001)

The BRI corresponds to the abundance-weighted average pollution tolerance of species occurring in a sample, and is similar to the weighted-average approach used in gradient analysis (Goff & Cottam, 1967; Gauch, 1982). The algorithm is:

$$I_s = \frac{\sum_{i=1}^{n} p_i \sqrt[3]{a_{si}}}{\sum_{i=1}^{n} \sqrt[3]{a_{si}}}$$

where I_s is the index value for sample s, n the number of species for sample s, p_i the position for species i on the pollution gradient (pollution-tolerance score) and a_{si} the abundance of species i in sample s.

According to the authors, determining the pollution-tolerance score (p_i) for the different species involves four steps: (1) assembling a calibration infaunal data set; (2) carrying out an ordination analysis to place each sample in the calibration set on a pollution gradient; (3) computing the average position of each species along the gradient and (4) standardising and scaling the position to achieve comparability across depth zones.

The average position of species I (p_i) on the pollution gradient defined in the ordination is computed as:

$$P_i = \frac{\sum_{j=1}^{t} g_j}{t}$$

where t is the number of samples to be used in the sum in which only the highest t species abundance values are included and g_j is the position on the pollution gradient in the ordination space for sample j.

This index has only been applied for assessing benthic infaunal communities on the Mayland Shelf of Southern California employing a 717-sample calibration data set.

2.3.1.1.5 BENTIX Index (Simboura & Zenetos, 2002)   The BENTIX Index was based on the *AMBI* Index (Borja *et al.*, 2000) and relies on the reduction of macrozoobenthic data from soft-bottom substrata in three wider ecological groups. To accomplish this goal, a list of indicators species was elaborated, where each species received a score, from 1 to 3, that represented their ecological group. In the light of the above the groups can be described as:

Group 1 (*GI*): includes the species that are sensitive or indifferent to disturbances (*K*-strategies species);
Group 2 (*GII*): includes the species that are tolerant and may increase their densities in case of disturbances, as well as the second-order opportunistic species (*r*-strategies species);
Group 3 (*GIII*): includes the first-order opportunistic species.

The formula that expresses this index is given by:

$$\text{BENTIX} = \left\{ \frac{6 \times \%GI + 2 \times (\%GII + \%GIII)}{100} \right\}$$

This index can range from 2 (Poor conditions) to 6 (High ecological quality status (EcoQs) or reference sites) (Table 3).

Overall, the BENTIX Index considers two major classes of organisms: the sensitive and the tolerant groups. This classification has the advantage of reducing the calculation effort while diminishing the probability of the inclusion of species in inadequate groups (Simboura & Zenetos, 2002). Moreover when using this index, it does not require for amphipoda identification expertise, since it encloses

Table 3: Categories considered as a function
of BENTIX Index values.

Classification	BENTIX value
Normal	4.5–6.0
Slightly polluted	3.5–4.5
Moderately polluted	2.5–3.5
Heavily polluted	2.0–2.5
Azoic	0

all those organisms (with exception to individuals from the *Jassa* genus) in the same category of sensitivity to organic matter increasing (Dauvin & Ruellet, 2007).

The BENTIX Index was developed in the scope of the WFD for the Mediterranean Sea. It has been successfully applied to cases of organic pollution (Simboura & Zenetos, 2002; Simboura *et al.*, 2005), oil spills (Zenetos *et al.*, 2004) and dumping of particulate metalliferous waste (Simboura *et al.*, 2007). This index is considered an ecologically relevant biotic index since it does not under or overestimates the role of any of the groups (Simboura *et al.*, 2005). Nevertheless, according to some authors, the BENTIX Index relies solely on the classification of organisms for organic pollution, being unable to accurately classify sites with toxic contaminations (Marín-Guirao *et al.*, 2005). It also emphasized the small lists of species, especially crustaceans, included in the attribution of the scores. Another point is that this index presents some limitations when applied to estuaries and lagoons (Simboura & Zenetos, 2002; Blanchet *et al.*, 2008).

2.3.1.1.6 Conservation Index (Moreno *et al.*, 2001) The Conservation Index is based on the health status of one marine seagrass, *P. oceanica.*

$$CI = \frac{L}{L + D}$$

Considering a given area under assessment, L is the proportion of living *P. oceanica* meadow and D the proportion of dead meadow coverage.

Different authors applied this index in areas near chemical industries, with results leading to establish four grades of *Posidonia* meadow conservation. These grades correspond to increasingly impacted areas, allowing the detection of changes in the industry activity as a function of the conservation status at a given location: <0.33 advanced regression; 0.33–0.56 impacted meadow; 0.56–0.79 low to moderate impact; >0.79 high conservation status.

2.3.1.1.7 EEI (Orfanidis *et al.*, 2001) Shifts in the structure and function of marine ecosystems are evaluated by classifying marine benthic macroalgae in two ecological groups (ESG I and ESG II). ESG I includes seaweed species with a thick or calcareous thalus, low growth rates and long life cycles, whereas the ESG II includes sheet-like and filamentous seaweed species with high growth rates and short life cycles.

The absolute abundance (%) of each ESG is estimated by coverage (%) in each sample. Obtaining at least three samples per season is recommended. The estimation of the EEI values and the equivalent ecological status is shown in Table 4.

2.3.1.1.8 Indicator Species Index (ISI) (Rygg, 2002) The ISI, which is based on the improved version of the Hulbert Index (1971), focuses on the assumption that each species reacts differently to pollution impacts, and consequently to the degradation of the ecosystem conditions. Knowing the species sensitivity to pollution factors, their presence or absence can be used to calculate the ISI in each

Table 4: Ecological Evaluation Index (EEI) values and equivalent ecological status (ESC), based on % coverage of marine benthic macroalgae from Ecological Status Group (ESG) I and II.

Mean coverage of ESG I (%)	Mean coverage of ESG II (%)	ESC	EEI	Spatial scale weighted EEI and equivalent ESCs
0–30	0–30	Moderate	6	≤6 to >4 = Moderate
	>30–60	Low	4	≤4 to >2 = Low
	>60	Bad	2	2 = Bad
>30–60	0–30	Moderate	8	≤8 to >6 = Good
	>30–60	Low	6	≤6 to >4 = Moderate
	>60	Bad	4	≤4 to >2 = Low
>60	0–30	Moderate	10	≤10 to >8 = High
	>30–60	Low	8	≤8 to >6 = Good
	>60	Bad	6	≤6 to >4 = Moderate

sample (it does not enter with taxa abundance). To calculate this index it is necessary to determine the sensitive values for each species as well as the pollution impact factor (ES100min5). The ES100 is the expected number of species among 100 individuals. The average of the five lowest ES100 was defined as the sensitivity value of that taxon, denoted ES100min5. The ISI is then defined as the average of the sensitivity values of the taxa occurring in the sample. This index allows an accurate description of the environmental quality of the systems and has been applied mostly in the Norway coasts.

One main disadvantage of the ISI is that it may not be transposed to other geographical regions without restrictions, since the taxonomic list can be significantly different and the calculation of the sensitivity factors may require different approaches. It has only been applied in the coast of Norway.

2.3.1.1.9   Macrofauna Monitoring Index (Roberts et al., 1998)   The main goal of the Macrofauna Monitoring Index (MMI) is assessing the impact of dredge oil dumping, based on monitoring indicator species in the benthic macrofauna. It is based on 12 indicator species, according to the criteria of easiness of identification, easy extraction from samples and representativeness. Each species is rated from 1 to 10 (very intolerant to impacts) based on the density of species in control versus impacted sites. A score of zero indicates a species, which is more common at impacted samples than at unimpacted sites. This score reflects basically the impacts that dredge spoil dumping have on its abundance. The index final score (is obtained by the averaged sum of all the species scores. Index values of <2.2, from 2.2 to 6 and >6 indicate severe impact, patchy impact and no impact, respectively. Its aim was the development of a site-specific monitoring index that would be statistically precise, biologically meaningful and very cost effective.

According to Roberts *et al.* (1998), the impacted sites show a higher content in mud or fine sand presenting lower macrofaunal abundance, diversity and richness than the unimpacted regions. This index identifies and estimates the stress and disturbance on the study site without appealing to exhaustive identification methods, since it relies on a small but informative subset of fauna. The MMI presents two main disadvantages: this index is a semi-quantitative measure of the degree of impact on macrofauna, correlating strongly with macrofaunal richness and abundance (Roberts *et al.*, 1998) and it is site- and pollution-type specific (Simboura & Zenetos, 2002).

2.3.1.1.10 Opportunistic macroalgae assessment tool (Scanlan *et al.*, 2007) The methodology underlying the approach taken for the development of a tool directed to monitor mats of various bloom-forming macroalgae on intertidal sedimentary shores was described by Scanlan *et al.* (2007), and in this case, the main pressure considered was eutrophication.

A selection of four basic tools was chosen based on those parameters that would describe or indicate the response of macroalgae to disturbance: Total available intertidal area for opportunistic macroalgae growth (ha); areal coverage (ha) – the total number of hectares of intertidal area effectively covered by macroalgae; percentage (%) cover (calculated according to Wither, 2003); and biomass (g WW m^{-2}).

The Opportunistic Macroalgae Assessment Tool combines, in a first step, the percentage cover with biomass to obtain a classification (see Tables 5 and 6).

For further details regarding boundary conditions, please address the mentioned paper. Secondly, to account for overall water-body size, the authors proposed that areal coverage (in hectares) should lower the class of a water body, as derived from Table 7, by one or more classes depending on the total area of algal mats (Scanlan *et al.*, 2007).

2.3.1.1.11 Pollution Index (Bellan-Santini, 1980) This index follows the same formulation and interpretation as the Annelida Pollution Index, but includes the amphipod group.

$$PI = \sum \frac{\text{Dominance of pollution indicators}}{\text{Dominance of clean water indicators}}$$

In this case, *Caprella acutrifans* and *Podocerus variegatus* are the pollution indicators and *Hyale* sp., *Elasmopus pocillamanus* and *Caprella liparotensis* the clear water indicators.

2.3.1.2 Bioaccumulator indicator species
There are species classified as bioaccumulative, defined as those capable of resisting and accumulating various pollutant substances in their tissues, which facilitates their detection whenever they are in the environment in very low

Table 5: Decision table for classification according to biomass and percentage cover (adapted from Scanlan et al., 2007).

Biomass (g WW m^{-2})					
>3000	M	M/P (entrained algae – monitor)	P	P	B
1000–3000	G/M (entrained algae – monitor)	M	M/P (entrained algae – monitor)	P	B
500–1000	G	G/M (entrained algae – monitor)	M	M	P
100 – 500	H	H/G (entrained algae – monitor)	G	G/M (entrained algae – monitor)	M
<100	H	H	G/M (entrained algae – monitor)	M	M/P (entrained algae – monitor)
% Cover	≤5	>5–15	>15–25	>25–75	>75–100

Note: Quality status: H, High; G, Good; M, Moderate; P, Poor and B, Bad.

Table 6: Decision table for classification according to percentage cover (from Scanlan et al., 2007).

	Classification				
	High	**Good**	**Moderate**	**Poor**	**Bad**
% Cover	<5%	5–15%	15–25%	25–75%	>75%

Table 7: Effect of total patch size on water-body classification class (from Scanlan et al., 2007).

Areal coverage	Effect on classification class
<100 ha	No change
100–499 ha	No change
500–999 ha	Downgrade by one class
1000–2499 ha	Downgrade by two classes
>2500 ha	Downgrade by three classes

levels which is otherwise difficult to detect through analytical techniques (Philips, 1977).

The disadvantage of using accumulator indicator species in the detection of pollutants arises from the fact that a number of biotic and abiotic variables may affect the rate at which the pollutant is accumulated, and therefore both laboratory and field tests need to be undertaken so that the effects of extraneous parameters can be identified.

Due to their sessile nature, wide geographical distribution and capability to accumulate toxic substances in their tissues and to detoxify when pollution ceases, the mollusc group, particularly bivalves, are the most used to determine the existence and quantity of toxic substances. Individuals of the genera *Mytilus* (De Wolf, 1975; Goldberg *et al.*, 1978; Dabbas *et al.*, 1984; Cossa & Rondeau, 1985; Miller, 1986; Renberg *et al.*, 1986; Carell *et al.*, 1987; Lauenstein *et al.*, 1990; Viarengo & Canesi, 1991; Regoli & Orlando, 1993), *Cerastoderma* (Riisgard *et al.*, 1985; Mohlenberg & Riisgard, 1988; Brock, 1992), *Ostrea* (Lauenstein *et al.*, 1990; Mo & Neilson, 1991) and *Donax* (Marina & Enzo, 1983; Romeo & Gnassia-Barelli, 1988) have been considered, in many works, as ideal in the detection of toxic substance concentration in the environment. In that sense, Goldberg *et al.* (1978) introduced the concept of "Mussel Watch" when referring to the use of molluscs in the detection of polluting substances. The National Oceanic and Atmospheric Agency (NOAA) in the United States of America has developed since 1980 the 'Mussel Watch Program', focused on pollution control along the North American coasts. Similar programmes exist in Canada (Cossa *et al.*, 1983; Picard-Berube & Cossa, 1983),

Denmark (Jensen *et al.*, 1981), the Mediterranean Sea (Leonzio *et al.*, 1981; Niencheski, 1982) and the North Sea (Golovenko *et al.*, 1981) and in the Australian coasts (Cooper *et al.*, 1982; Ritz *et al.*, 1982; Wooton & Lye, 1982; Richardson & Waid, 1983).

Similarly, certain amphipod species are considered capable of accumulating toxic substances (Albrecht *et al.*, 1981; Reish, 1993), as well as polychaete species like *Nereis diversicolor* (Langston *et al.*, 1987; McElroy, 1988), *Neanthes arenaceodentata* (Reish & Gerlinger, 1984), *Glycera alba, Tharix marioni* (Gibbs *et al.*, 1983) or *Nephtys hombergi* (Bryan & Gibbs, 1987). Some fish species have also been used in various works focused on the effects of toxic pollution of the marine environment, due to their bioaccumulative capability (Eadie *et al.*, 1982; Gosset *et al.*, 1983; Varanasi *et al.*, 1989) and the existing relationship between pathologies suffered by benthic fish and the presence of polluting substances (Malins *et al.*, 1984; Couch & Harshbarger, 1985; Myers *et al.*, 1987).

Other authors such as Levine (1984), Maeda & Sakaguchi (1990), Neumann *et al.* (1991) and Storelli & Marcotrigiano (2001) view algae as the most favourable detectors of heavy metals, pesticides and radionuclides, where *Fucus, Ascophyllum* and *Ulva* are the most utilized taxa.

2.3.1.2.1 The Ecological Reference Index (ERI) For reasons of comparison, the concentration of substances in organisms must be translated to uniform and comparable units. This is done through the ERI, which represents a potential for environmental effects. This index has only been applied using blue mussels.

$$\text{ERI} = \frac{\text{measured concentration}}{BCR}$$

BCR stands for the value of the background/reference concentration. The upper limit of *BCR* for hazardous substances in blue mussels according to OSPAR/MON (1998) is provided in Table 8.

Table 8: Upper limits of *BCR* for hazardous substances in blue mussels (OSPAR/MON, 1998).

Substance	Upper limit of *BCR* value (ng g^{-1} dry weight)
Cadmium	550
Mercury	50
Lead	959
Zinc	150,000

Few indices like the ERI, based on the use of bioaccumulative species, have been proposed. It is in fact more common to simply measure the effects (e.g. incidence and mortality percentage) of a certain pollutant on those species, or to use biomarkers, which can be useful in evaluating the specificity of responses to natural or anthropogenic changes. Nevertheless, it is very difficult for environmental managers to interpret increasing or decreasing changes in biomarker data.

The Working Group on Biological Effects of Contaminants (WGBEC, 2002) recommended different techniques for biological monitoring programmes, which are summarized in Table 9.

2.3.2 Indicators based on ecological strategies

The intention of some indicators is to assess environmental-stress effects taking the ecological strategies followed by different organisms into consideration. That is the case of trophic indices such as the Infaunal Trophic Index (ITI) (Word, 1979) and the Feeding Structure Index (FSI), based on organisms' different feeding strategies. Another example is the Nematodes/Copepods Index (Raffaelli & Mason, 1981) that accounts for the different behaviour of two taxonomic groups under environmental-stress situations. Nevertheless, several authors rejected these types of indicators due to their dependence on parameters like depth and sediment particle size, as well as because of their unpredictable pattern of variation depending on the type of pollution (Gee et al., 1985; Lambshead & Platt, 1985).

Other proposals such as the Meiobenthic Pollution Index (Losovskaya, 1983), the Mollusc Mortality Index (Petrov, 1990), the Indice of Trophic Diversity (ITD) (Heip et al., 1985), the Maturity Index (MI) (Bongers et al., 1991), the Polychaeta Amphipoda Ratio (Gómez-Gesteira & Dauvin, 2000) revised as Benthic Opportunistic Polychaeta Amphipoda (BOPA) (Dauvin & Ruellet, 2007), or the Index of r/K strategies proposed by De Boer et al. (2001) appeared.

The Rhodophyceae/Phaeophyceae Index proposed by Feldmann (1937), based on marine vegetation, is widely used in the Mediterranean Sea. It was established as a biogeographical index and accounts for the fact that the number of Rodophyceae species decreases from the tropics to the poles. Its application as indicator holds on the higher or lower sensitivity of Phaeophyceae and Rhodophyceae to disturbances. In addition, Belsher (1982) proposed an index based on the qualitative and quantitative dominance of each taxonomic group.

Below are listed the indices based on ecological strategies most commonly used in assessing coastal and marine environments.

2.3.2.1 Belsher Index (Belsher, 1982)
This index, elaborated in 1982 by T. Belsher, holds on the higher or lower sensitivity of Phaeophyceae and Rhodophyceae to disturbances and is based on the qualitative and quantitative dominance of each taxonomic group.

Table 9: Review of different techniques for biological monitoring.

Method	Organism	Issues addressed	Biological significance	Threshold value
Bulky DNA adduct formation	Fish	PAHs, Other synthetic organics	Measures: genotoxic effects. Sensitive indicator of past and present exposure	2 × Reference site or 20% change
AchE	Fish and Bivalve molluscs	Organophosphates and carbonates or similar molecules.	Measures: exposure	Minus 2.5 × Reference site
Metallothionein induction	Fish and *Mytilus* sp.	Measures: induction of metallothionein protein by certain metals	Measures: exposure and disturbance of copper and zinc metabolism.	2.0 × Reference site
EROD or P4501A induction	Fish	Measures: induction of enzymes with metabolized planar organic contaminants		2.5 × Reference site
ALA-D inhibition	Fish	Lead	Index of exposure	2.0 × Reference site
PAH bile metabolites	Fish	PAHs	Measures: exposure to and metabolism PAHs	2.0 × Reference site
Lysosomal stability	Fish and *Mytilus* sp.	Not contaminant specific but responds to a wide variety of xenobiotics, other contaminants and metals	Provides a link between exposure and pathological endpoints	2.5 × Reference site
Lysosomal neutral red retention	*Mytilus* sp.	Not contaminant specific but responds to a wide variety of xenobiotics, other contaminants and metals	Provides a link between exposure and pathological endpoints	2.5 × Reference site

(Continued)

Table 9: (Continued)

Method	Organism	Issues addressed	Biological significance	Threshold value
Early toxicopathic lesions, pre-neoplastic and neoplastic liver histopathology	Fish	PAHs	Measures: pathological changes associated with exposure to genotoxic and non-genotoxic carcinogens	2.0 × Reference site or 20% change
Scope for growth	Bivalve molluscs	Responds to a wide variety of contaminants	Integrative response which is a sensitive and sublethal measure of energy available for growth	
Shell thickening	*Crassostea gigas*	Specific to organotins	Disruption to pattern of shell growth	
Vitellogenin induction	Male and juvenile fish	Oestrogenic substances	Measures: feminization of male fish and reproductive impairment	
Imposex	Neogastropod molluscs	Specific to organotins	Reproductive interference	2.0 × Reference site or 20% change
Intersex	*Littorina littorina*	Specific to reproductive effects of organotins	Reproductive interference	2.0 × Reference site or 20% change
Reproductive success in fish	*Zoarces viviparus*	Not contaminant specific	Measures: reproductive output and survival of eggs and fry in relation to contaminants	

$$\text{Qualitative Dominance} = \frac{\% \text{ species of a taxonomic group}}{\sum \text{population species}} \times 100$$

$$\text{Quantitative Dominance} = \frac{\% \text{ cover area by a group}}{\text{total cover area}}$$

The ratio between qualitative and quantitative dominance is called tension Ψ. It has been observed that alongside decreasing pollution gradients, certain groups of algae increase or decrease their tension, establishing the following relation, considered a Pollution Index:

$$\frac{\sum \psi i}{\sum \psi j}$$

where i = groups with decreasing tension and j = groups with increasing tension.

The Pollution Index values are high in polluted areas and nearly null in uncontaminated zones, but boundary conditions between different states of disturbance are not evident. This means that the knowledge about this index's behaviour does not seem to be enough to consider it a good pollution indicator. It has only been applied in rocky substrate areas.

2.3.2.2 Benthic Opportunistic Polychaeta Amphipoda Index, BOPA (Dauvin & Ruellet, 2007)

The BOPA Index results from the refinement of the Polychaeta/Amphipoda Index (Gómez-Gesteira & Dauvin, 2000), to be applicable under the WFD perspective. Accordingly, this index can be used to assign estuarine and coastal communities into five EcoQs categories. This index, in accordance with the taxonomic-sufficiency principle, aims to exploit this ratio to determine the ecological quality, using relative frequencies ([0;1]) rather than abundances ([0;+∞]) to define the limits of the index. This way, it can be written as:

$$\text{BOPA} = \text{Log}\left\{ \frac{f_P}{f_A + 1} + 1 \right\}$$

where f_P is the opportunistic polychaeta frequency (ratio of the total number of opportunistic polychaeta individuals to the total number of individuals in the sample); f_A the amphipoda frequency (ratio of the total number of amphipoda individuals, excluding the opportunistic *Jassa* sp. amphipod, to the total number of individuals in the sample), and $f_P + f_A \leq 1$. Its value can range between 0 (when $f_P = 0$) and Log 2 (around 0.30103, when $f_A = 0$). The BOPA Index will get a null value only when there are no opportunistic polychaetes, indicating an area with a very low amount of organic matter. So, when the index presents low values it is considered that the area has a good environmental quality, with few opportunistic species; and it increases as increasing organic matter degrades the environmental conditions.

One of the main advantages of this index is its independence from sampling protocols, and specifically of mesh sieve sizes, since it uses frequency data and the proportion of each category of organisms. The need for taxonomic knowledge is reduced, which allows a generalized use and ease of implementation. Moreover, the use of frequencies makes it independent of the surface unit chosen to express abundances and it is sensitive to increasing organic matter in sediment as well as to oil pollution. Nevertheless, it takes into account only three categories of organisms – opportunistic polychaetes, amphipods (except *Jassa* sp.) and other species – but only the first two have a direct effect on the index calculation. Another point is that it does not consider the Oligochaeta influence, which may also include opportunistic species.

2.3.2.3   FSI (Milovidova & Alyomov, 1992)

$$I = \frac{N \text{ species of filter-feeders}}{N \text{ species of deposit-feeders} + \text{predators}}$$

This index is based on the fact that in less-eutrophied areas, the number of filter-feeder species is six to eight times greater than in highly eutrophied areas (Petrov & Shadrina, 1996). Its application is complex due to the difficulty in correctly assigning a trophic category to each individual.

2.3.2.4   ITI (Word, 1980)

Macrozoobenthic species can be divided in: (1) suspension feeders that collect detrital materials in overlying water using appendages of the animal or tube or burrow capturing strategies where currents settle these materials adjacent to the organisms; (2) interface feeders that collect detrital materials that settle on the surface of the sediment – ingested particles are generally less than 50 microns in diameter; (3) surface-deposit feeders that collect larger particles that are contained within the upper 2 cm of the sediment layer and (4) subsurface deposit feeders that generally collect particles that are buried deeper than 2 cm. Specialised feeders of this last guild also include species that use methane as a food source. The index value is given by:

$$ITI = 100 - \frac{100}{3} \times \frac{(0n_1 + 1n_2 + 2n_3 + 3n_4)}{(n_1 + n_2 + n_3 + n_4)}$$

in which n_1, n_2, n_3 and n_4 are the number of individuals sampled in each of the above-mentioned groups.

ITI values near 100 mean that suspension feeders are dominant and the environment is not disturbed. At values near 0, subsurface feeders are dominant, meaning that the environment is strongly disturbed, probably due to human activities. Index values less than 60 are highly correlated to BOD and TOC or volatile solids in the upper 2 cm of the sediment, while values above 60 are less correlated to accumulation of organic materials in the sediment (Word, 1990).

2.3.2.5   ITD (Heip *et al.*, 1985)

Heip *et al.* (1985) proposed the ITD based on Nematodes trophic guilds. Nematodes are particularly suitable for the identification of their trophic habits, because their buccal structures allow a relatively easy and widely accepted classification into trophic guilds (Wieser, 1953; Jensen, 1987; Moens & Vincx, 1997; Moens *et al.*, 1999). Nematodes must be divided into four original groupings as follows (Wieser, 1953):

(1A) no buccal cavity or a fine tubular one – selective deposit (bacterial) feeders;
(1B) large but unarmed buccal cavity – non-selective deposit feeders;
(2A) buccal cavity with scraping tooth or teeth – epistrate (diatom) feeders;
(2B) buccal cavity with large jaws – predators/omnivores.

The ITD is then calculated as follows: ITD$=\Sigma\theta^2$, where θ is the contribution of density of each trophic group to total nematode density (Heip *et al.*, 1985). ITD ranges from 0.25 (highest trophic diversity, i.e. the four trophic guilds account for 25% each) to 1.0 (lowest diversity, i.e. one trophic guild accounts for 100% of nematode density).

2.3.2.6   MI (Bongers *et al.*, 1991; Bongers, 1999)

Nematodes are increasingly being used in environmental studies. One of the potential parameters to measure the impact of disturbances and to monitor changes in the structure and functioning of the below-ground ecosystem is the Nematode Maturity Index; an index based on the proportion of colonizers (*r*-strategists) and persisters (*K*-strategists) in samples (Bongers, 1999). This index my be applied in terrestrial, freshwater, marine and brackish habitats.

The colonizer–persister scale, the basis for the MI, is composed of five classes, 1–5; the colonizers, characterized by a high reproduction receive a low value, the persisters, which reproduce slowly, are allocated to cp–5. The MI is calculated as the weighted mean of the individual taxon scores:

$$MI = \sum_{i=1}^{n} v(i).f(i)$$

where $v(i)$ = the c–p value of taxon I as given in (Bongers *et al.*, 1991) and $f(i)$ = the frequency of that taxon.

The MI in practice varies from 1, under extremely enriched conditions, to a value between 3 or 4 under undisturbed conditions. Bongers *et al.* (1991) tested the MI behaviour in several conditions (organic pollution, oil spill, heavy metals contamination and recolonization). The case studies suggest that the MI is decreased by pollution (sewage waste, oil, heavy metals) but increases during the colonization process. Both disturbance and an increase in decomposition rate can be seen to result in a decreasing MI.

2.3.2.7   Meiobenthic Pollution Index (MMI) (Losovskaya, 1983)

$$\text{MPI} = \frac{\lg(H+1) + \lg(P+1)}{2\lg N}$$

where H, P and N are the number (ind m^{-2}) of Harpacticoida, Polychaeta and Nematoda, respectively, in a given benthic sample.

Increasing impacts induce the replacement of harpacticoides and polychaetes by nematodes, and such a shift can be traced through changes in the values of the index.

2.3.2.8   Mollusc Mortality Index (Petrov, 1990)

$$\text{MMI}(\%) = \frac{\text{Weight of shells of recently dead molluscs}}{\text{Total weight of living individuals and the shells of molluscs of the same species}}$$

High values of the index are indicative of disturbances.

2.3.2.9   Nematodes/Copepods Index (Raffaelli & Mason, 1981)

This index is based on the ratio between the abundances of nematodes and copepods.

$$I = \frac{\text{Nematodes abundance}}{\text{Copepodes abundance}}$$

The values of such a ratio can increase or decrease in response to higher or lower organic pollution, which expresses a different response of those groups to the input of organic matter into the system. Values over 100 express high organic pollution.

According to different authors, the application of this index should be limited to intertidal areas, since in infralittoral zones, at given depths and despite the absence of pollution, values observed were very high. This fact is explained by the absence of copepods at such depths, most probably due to a change in the optimal interstitial habitat for that taxonomic group (Krogh & Spark, 1936 and Wigely & Mcintyre, 1964 in Raffaelli & Mason, 1981).

2.3.2.10   Polychaeta/Amphipoda Ratio (Gómez-Gesteira & Dauvin, 2000)

This index follows similar principles as the Nematodes/Copepods Index, but it is applied to the macrofaunal level using polychaetes and amphipods. The index was formerly intended to measure the effects of pollution by crude.

$$I = \text{Log}_{10}\left(\frac{\text{Polychaete abundance}}{\text{Amphipod abundance}} + 1\right)$$

$I \leq 1$: non-polluted
$I > 1$: polluted

2.3.2.11   Rhodophyceae/Phaeophyceae Index (Feldmann, 1937)

The biogeographical Rhodophyceae/Phaeophyceae Index (Feldmann, 1937) has been proposed to assess the effects of environmental pollution on macroalgal communities and is given by:

$$I = \frac{\text{Number of Rhodophyceae species}}{\text{Number of Phaeophyceae species}}$$

Cormaci & Furnari (1991) detected values over 8 for this index in polluted areas in Southern Italy when the normal values in a well-balanced community vary between 2.5 and 4.5. Verlaque (1977) studied the effects of a thermal power station, and also found higher values, although this author viewed such results as being due to the presence of communities of warmth affinity. However, Belsher & Boudouresque (1976) analysed the submersed vegetation in small harbours and found that in such conditions the Phaeophyceae show higher proliferation, which decreases the index value. Therefore, the knowledge about this index's behaviour does not seem to be enough to regard it as a good pollution indicator in itself.

2.3.3   Indicators based on diversity

Diversity is one of the most-used concepts in assessing pollution, supported by the fact that the relationship between diversity and environmental disturbances can be seen as an inverse one. Magurran (1989) divides diversity measurements into three main categories: (1) Indices that measure species richness, such as the Margalef Index, which are essentially measurements of the number of species in a defined sampling unit; (2) Models of species abundance, such as the K-dominance curves (Lambshead et al., 1983) or the log-normal model (Gray, 1979), which describe the distribution of their abundances, ranging from those that represent situations in which there is a high uniformity to those that characterise cases in which the abundance of each species is very unequal. It must be said that the log-normal model deviation was rejected by several authors since it was impossible to find any benthic marine sample that clearly responded to such a distribution model (Shaw et al., 1983; Hughes, 1984; Lambshead & Platt, 1985) and (3) Indices based on the proportional abundance of different species, which intend to account for richness and uniformity in a simple expression. This category of indices can also be divided into those based on (1) statistics, (2) Information Theory and (3) species dominance. Indices derived from the Information Theory, such as the Shannon–Wiener, are based on something logical, such as diversity. Information in a natural system can be measured in a similar way as information contained in a code or message. On the other hand, dominance indices such as the Simpson or Berger–Parker are referred as measurements that ponder the abundance of the most-common species, instead of species richness.

In the meantime, Taxonomic Distinctness measures have been proposed and used by some researchers (e.g. Warwick & Clarke, 1995, 1998; Clarke &

Warwick, 1999) to evaluate biodiversity in the marine environment, taking into account taxonomical, numerical, ecological, genetic and phylogenetic aspects of diversity. These measures address some of the problems identified in relation to species richness and other diversity indices (Warwick & Clarke, 1995).

The most commonly used diversity measures are listed below.

2.3.3.1   Berger–Parker Index (Berger & Parker, 1970; May, 1975)

This index expresses the proportional importance of the most-abundant species, and may be computed using the following algorithm:

$$D = \frac{n_{max}}{N}$$

where n_{max} is the number of individuals of the most-abundant species and N the total number of individuals. The index values may vary from 0 to 1 and, contrary to other diversity indices, higher values correspond to a lower diversity.

2.3.3.2   Deviation from the log-normal distribution (Gray & Mirza, 1979)

This method, proposed by Gray & Mirza (1979), is based on the assumption that when a sample is taken from a community, the individuals' distribution tends to follow a log-normal model. The adjustment to a logarithmic normal distribution assumes that the population is ruled by a certain number of factors and is at a steady equilibrium. Consequently, any deviation from such distribution implies that some perturbation is affecting it.

2.3.3.3   Fisher's α Index (Fisher et al., 1943)

$$S = \alpha \times \ln\left(1 + \frac{n}{\alpha}\right)$$

where S is the number of taxa, n the number of individuals and α the Fisher's α, which is the shape parameter, fitted by maximum likelihood, under the assumption that the distribution of species abundance follows a log series. This has certainly been shown to be the case with some ecological data sets, but can by no means be universally assumed, and its use is restricted to genuine (integral) counts (Warwick & Clarke, 2001).

2.3.3.4   Hulbert Index (Hulbert, 1971)

$$ESn = \sum_{i=1}^{n}\left[1 - \frac{(N - Ni)!(N - n)}{(N - Ni - n)!N!}\right]$$

where N is the total number of individuals in a sample and Ni is the number of individuals of the i-th species.

The idea is to generate an absolute measure of species richness, which can be compared across samples of extremely different sizes. Nevertheless, the validity of this index depends on the assumption that the individuals of each species are randomly distributed, which is not always the case.

2.3.3.5   *K*-Dominance curves (Lambshead *et al.*, 1983)

The *K*-dominance curve is the representation of the accumulated percentage of abundance versus the logarithm of the sequence of species ranked in decreasing order. The slope of the straight line obtained allows the valuation of the pollution grade. The higher the slope, the higher the diversity will be.

2.3.3.6   Margalef Index (Margalef, 1969)

The Margalef Index quantifies diversity by relating specific richness to the total number of individuals.

$$D = \frac{(S-1)}{\log_e N}$$

where S = number of species and N = total number of individuals.

The author did not establish any reference values, and in fact the main problem when applying this index is the absence of a limit value, and therefore the difficulty in establishing such reference values. Ros & Cardell (1991) consider values below 4 as typical of polluted areas. On the other hand, Bellan–Santini (1980) settled a different limit, considering an area polluted when the index takes on values below 2.05.

2.3.3.7   Pielou Evenness Index (Pielou, 1969)

This index, introduced by Pielou in 1969, is a measure of how evenly distributed abundance is among the species that exist in a community.

$$J' = \frac{H'}{H'_{max}} = \frac{H'}{\log S}$$

where H'_{max} is the maximum possible value of the Shannon diversity.

The values of this index may vary from 0 to 1, where 1 represents a community with perfect evenness, and decreases to zero as the relative abundances of the species diverge from evenness.

2.3.3.8   Ranked species abundance (dominance) curves

It consists of ranking the species (or higher taxa) in decreasing order of their importance in terms of abundance or biomass. The ranked abundances, expressed as a percentage of the total abundance of all species, are plotted against the relevant species rank.

2.3.3.9   Rarefaction curves (Sanders, 1968)

Rarefaction curves are plots of the number of individuals on the *x*-axis against the number of species on the *y*-axis. The more diverse the community, the steeper and more elevated the rarefaction curve is.

2.3.3.10   Shannon–Wiener Index (Shannon & Weaver, 1963)

This index is based on the Information Theory. It assumes that individuals are sampled at random, out of an "indefinitely large" community, and that all the species are represented in the sample and can be estimated according to the algorithm:

$$H' = -\sum p_i \log_2 p_i$$

where p_i is the proportion of individuals belonging to species i in the sample. The real value of p_i is unknown, but it is estimated through the ratio $N_i\ N$, where N_i = number of individuals of the species i and N = total number of individuals.

The units for the index depend on the log used. Thus, for $\log_2$, the unit is bits/individual; "natural bels" and "nat" for $\log_e$ and "decimal digits" and "decits" for $\log_{10}$.

The index can usually take values between 0 and 5, and maximal values above 5 bits/individual are very rare. In this case, diversity is a logarithmic measurement, which makes it a fairly sensitive index in the range of values next to the upper limit (Margalef, 1978).

As a common basis, in the literature, low values are regarded as an indication of pollution (Stirn et al., 1971; Anger, 1975; Hong, 1983; Zabala et al., 1983; Encalada & Millan, 1990; Calderón-Aguilera, 1992; Pocklington et al., 1994; Engle et al., 1994, Mendez-Ubach, 1997; Yokoyama, 1997). But one of the problems arising with its use is the lack of objectivity, when trying to establish in a precise manner, from what threshold one should start taking into account the index values as an indication of the effects of such pollution.

For instance, Molvær et al. (1997) established the following relation between the Shannon–Wiener Index values and the different levels of ecological quality (Table 10), in accordance with thatrecommended by the WFD (WFD, 2000/60/CE).

Table 10:   Categories considered as a function of Shannon–Wiener Index values, according to Molvær et al. (1997).

Classification	Shannon–Wiener value
High status	>4 bits/individual
Good status	4–3 bits/individual
Moderate status	3–2 bits/ individual
Poor status	2–1 bits/ individual
Bad status	1–0 bits/ individual

Detractors of this index based their criticisms on its lack of sensitivity when having to detect the initial stages of pollution (Leppäkoski, 1975; Pearson & Rosenberg, 1978; Rygg, 1985). For instance, Gray (1979), studying the effects of a cellulose paste factory waste, pointed out the uselessness of this index since it responds to such obvious changes where there is no need for a detection tool.

Ros & Cardell (1991), in their study on the effects of great industrial and human domestic pollution, consider the index as a partial approach to the knowledge of pollution effects on marine benthic communities and, without any further explanation to that statement, set out a new structural index proposal, whose lack of applicability has already been shown by Salas (2002).

2.3.3.11   Simpson Index

Simpson (1949) proposed a diversity index that accounts for the probability of two individuals randomly sampled from an infinitely large community belonging to the same species:

$$D = \sum p_i^2$$

where p_i is the proportion of individuals from species i in the community.

To calculate the index for a finite community the following algorithm can be used:

$$D = \sum \left[\frac{n_i(n_i - 1)}{N(N - 1)} \right]$$

where n_i is the number of individuals of species i and N is the total number of individuals.

Similar to the Berger–Parker Index, the Simpson Index may vary from 0 to 1, it has no dimensions and, in the same way, higher values correspond to lower diversity.

2.3.3.12   Taxonomic distinctness measures (Warwick & Clarke, 1995; Clarke & Warwick, 1998, 2001)

To estimate Taxonomic Diversity indices, a hierarchical Linnean classification is used as a proxy for cladograms representing the relatedness of individual species. For each location, a composite taxonomy is compiled and five taxonomic levels are considered (species, genus, family, order, class and phylum). Several measures integrating information usually provided by species richness and other diversity indices were proposed. The measures proposed by R.M. Warwick and K.R. Clarke are provided below and may be calculated from macrofauna abundance, using PRIMER 6 (Software package from Plymouth Marine Laboratory, United Kingdom):

a) Taxonomic Diversity, Δ

$$\Delta = \frac{[\sum\sum_{i<j} w_{ij} x_i x_j]}{[n(n-1)/2]} \tag{1}$$

b) Taxonomic Distinctness, Δ^*

$$\Delta^* = \frac{[\sum\sum_{i<j} w_{ij} x_i x_j]}{[\sum\sum_{i<j} x_i x_j]} \tag{2}$$

c) Average Taxonomic Distinctness (based on presence/absence of species) Δ^+

$$\Delta^+ = \frac{[\sum\sum_{i<j} w_{ij}]}{[s(s-1)/2]} \tag{3}$$

d) Variation in Taxonomic Distinctness Λ^+

$$\Lambda^+ = \frac{[\sum\sum_{i\neq j}(w_{ij} - \bar{w})^2]}{[s(s-1)]} \tag{4}$$

e) Total Taxonomic Distinctness, $s\Delta^+$

$$s\Delta^+ = \sum_i \left| \frac{(\sum_{i\neq j} w_{ij})}{(s-1)} \right| \tag{5}$$

where x_i represents the abundance of the ith of s species observed, $n(=\sum_i x_i)$ is the total number of individuals in the sample and w_{ij} the 'distinctness weight' given to the path-length-linking species i and j in the taxonomy.

Taxonomic Diversity (eq. (1)) can be thought of as the average path length between two randomly chosen individuals from the sample (including individuals of the same species), whereas Taxonomic Distinctness (eq. (2)) is the average path length between two randomly chosen individuals, conditional on them being from different species (Rogers *et al.*, 1999). Using data consisting only of the presence or absence of species (i.e., species list), a simpler form of Taxonomic Distinctness eq. (3) can be thought of as the average length between any two randomly chosen species present in the sample. The degree to which taxa are over- or underrepresented in samples is another biodiversity attribute of ecological relevance and is reflected by the Variation in Taxonomic Distinctness (eq. (4)). Finally, Total Taxonomic Distinctness (eq. (5)) was proposed by Clarke & Warwick as a useful measure of total taxonomic breadth of an assemblage, as a modification of species richness, which allows for the species interrelatedness.

2.3.4   Indicators based on species biomass and abundance

Other approaches account for the variation of organism's biomass and abundance as a measure of environmental disturbances. These approaches encompass methods such as the SAB Curves (Pearson & Rosenberg, 1978), consisting of a comparison between the curves resulting from ranking the species as a function of their representativeness in terms of both their abundance and biomass. The use of this method is not advisable because it is purely graphical, which leads to a high degree of subjectivity and does not allow relating it quantitatively with the environmental factors. In its turn, the ABC Method (Warwick, 1986) also involves the comparison between the cumulative curves of species biomass and abundance. Warwick & Clarke (1994) derived the *W*-Statistic Index from this method.

2.3.4.1   ABC method (Warwick, 1986)

This method is based on the assumption that, for a given community, the distribution of the number of individuals and the biomass of each species does not show the same variation pattern. It consists, in fact, of an adaptation from the already mentioned *K*-dominance curves, although it shows the *K*-dominance and the biomass curves in a single graphic. The graphics allow plotting the interval of species (in the abscissa axis), arranged in decreasing order according to a logarithmic scale, against the cumulative dominance curves (in the ordinate axis).

Three different situations can occur as function of the degree of disturbance affecting the community (Figure 2):

1. In a non-disturbed system, a fairly low number of relatively large individuals of few species will contribute with most of the biomass, and at the same time, the

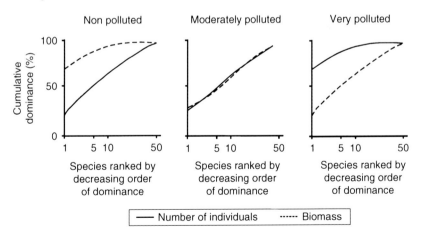

Figure 2:   Hypothetical *K*-dominance curves for species biomass and number, showing unpolluted, moderately polluted, and heavily polluted conditions. On the horizontal (*X*) axis, the species are ranked by decreasing order of importance, using a log scale. On the vertical (*Y*) axis, the percentage dominance is plotted using a cumulative percentage scale (Warwick, 1986).

individuals' distribution among the different species is more equitative. Graphically, the biomass curve will be plotted above the abundance one, indicating higher numeric diversity rather than biomass diversity.

2. In communities under moderate disturbance conditions, the biomass cumulative curve will not show such an important contribution of just a few species represented by a low number of individuals as in the previous case, but on the other hand, abundances increase. Graphically, the biomass and abundance curves come out intersected.

3. In the case of communities under intense disturbances, only a few species will represent most of the individuals, all of a small size, which explains why the biomass of each one of the species is low and more equitatively shared. Graphically, the abundance curve comes out above the biomass curve, indicating higher biomass rather than numerical diversity in the distributions.

2.3.4.2   *W*-Statistic Index (Warwick & Clarke, 1994)

Some authors, namely Beukema (1988), Clarke (1990), McManus & Pauly (1990) and Meire & Dereu (1990) have tried to convert the ABC Method into a measurable index. Clarke's (1990) approach became the most commonly accepted.

$$W = \sum_{i=1}^{s} \frac{(B_i - A_i)}{50(S-1)}$$

where B_i is the biomass of species i, A_i the abundance of specie i and S the number of species. The index can take values from +1 (indicating a non-disturbed system (High status)) to -1 (which defines a polluted situation (Bad status)). Values close to 0 indicate moderate pollution (Moderate status).

This approach is specific to organic pollution and has been applied, with satisfactory results, to soft-bottom tropical communities (Anderlini & Wear, 1992; Agard et al., 1993), to experiments (Gray et al., 1988), to fish found in disturbed areas (Ritz et al., 1989) and on coastal lagoons (Reizopoulou et al., 1996; Salas, 2002). However, Ibanez & Dauvin (1988), Beukema (1988), Weston (1990), Craeymeersch (1991) and Salas et al. (2004) obtained confusing results after applying this method to assess the environmental status of estuarine zones, which was induced by the appearance of dominant species in normal conditions, favoured not by organic pollution but by other environmental factors. On the other hand, in spite of having been designed to be applied to benthic macrofauna, Abou-Aisha et al. (1995) applied this method in three areas of the Red Sea to detect the impact of phosphorus wastes on macroalgae. Nevertheless, problems may arise when applying it to marine vegetation, due to obvious difficulties in counting the number of individuals of vegetal species.

2.3.5   Multimetric indices

From a more holistic point of view, some authors proposed indices capable of integrating different types of environmental information. A first approach was developed by Satsmadjis (1982) for application in coastal areas, relating sediment particle size to the diversity of benthic organisms. Jeffrey *et al.* (1985) developed the Pollution Load Index (PLI), Rhoads & Germamo (1986) proposed the Organism Sediment Index (OSI). Vollenweider *et al.* (1998) developed a TRIX integrating chlorophyll *a*, oxygen saturation, total nitrogen and phosphorus to characterise the trophic state of coastal waters. In the same way, Fano *et al.* (2003) proposed the Ecofunctional Quality Index (EQI) that considers the macrofaunal and macrophytic abundance/biomass. In a progressively more complex way, other indices such as the IBI for coastal systems (Nelson, 1990), the Benthic Condition Index (Engle *et al.*, 1994; Macauley *et al.*, 1999), the Benthic Habitat Quality (BHQ) (Nilsson & Rosenberg, 1997), the BRI (Smith *et al.*, 2001), the Biological Quality Index (BQI) (Rosenberg *et al.*, 2004) or the Chesapeake Bay B-IBI Index (Weisberg *et al.*, 1997), the Carolina Province B-IBI (Van Dolah *et al.*, 1999) and the Virginia Province Benthic Index (VPBI) (Paul *et al.*, 2001) include physicochemical factors, diversity measures, specific richness, taxonomical composition and the system's trophic structure.

Similarly, a set of specific indices of fish communities has been developed to measure the ecological status of estuarine areas. The Estuarine Biological Health Index (BHI) (McGinty & Lider, 1997) combines two separate measures (health and importance) into a single index. The Fish Health Index (FHI) (Cooper *et al.*, 1993) is based on both qualitative and quantitative comparisons with a reference fish community. The Estuarine Ecological Index (EBI) (Deegan *et al.*, 1993) reflects the relationship between anthropogenic alterations in the ecosystem and the status of higher trophic levels, and the Estuarine Fish Importance Rating (FIR) is based on a scoring system of seven criteria that reflect the potential importance of estuaries for the associated fish species. This index is able to provide a ranking, based on the importance of each estuary and helps to identify the systems with major importance for fish conservation.

Nevertheless, these indices are not often used in a generalized way because they have usually been developed to be applied in a particular system or area, which turns them dependent on the type of habitat and seasonality.

More recently, Hale & Heltshe (2008) developed the Acadian Province Benthic Index (APBI). With the challenges brought by the WFD implementation other multimetric indices were proposed, namely the Multivariate-AMBI (M-AMBI) (Muxika *et al.*, 2007) and the Portuguese Benthic Assessment Tool (P-BAT) (Teixeira *et al.*, submitted). Pinto *et al.* (2009) gives a detailed revision regarding the most-used biotic indices.

2.3.5.1   APBI (Hale & Heltshe, 2008)

This index resulted from the need to develop a multivariate tool for the Gulf of Maine. The intent was to use this index as an ecological indicator of benthic condition along the coast and for year-to-year comparisons. To achieve this point, environmental standards – called Benthic Environmental Quality scores (BEQ) – that would be used as reference conditions during the index development and performance were established. The APBI is based on each station

BEQ classifications that the best-candidate metrics were selected and tested and gave rise to the APBI development. The authors also considered the predictive value of an indicator, based on the function of its sensitivity, specificity and the prevalence of the condition it is supposed to indicate. The positive predictive value (PPV) is the probability of a positive response (low BEQ), given that the indicator is positive (low APBI). The negative predictive value (NPV) is the probability of negative response, given that the indicator is negative.

Being the Logit(p) function from multivariate logistic regression, the model that best fits the obtained data for that particular region and conditions, among a broad range of different combinations of benthic metrics, was the one given by:

$$\text{Logit}(p) = 6.13 - 0.76\,H' - 0.84\,\text{Mn\_ES}(50)._{.05} + 0.05\,\text{PctCapitellidae} \quad (1)$$

where H' is the Shannon–Wiener Index, with higher scores representing higher mean diversity; $\text{Mn\_ES}(50).05$ is the station mean of the Rosenberg $et\ al.$ (2004) species tolerance value (higher scores meaning more pollution sensitivity); Pct-Capitellidae is the percent abundance of Capitellidae polychaetes, once more, higher scores meaning more capitellids, which do well in organically enriched sediments. Based on this, the index probability can be computed as:

$$p = \frac{e^{(\text{Logit}(p))}}{1 + e^{(\text{Logit}(p))}}$$

where p is the probability that BEQ is low. A higher H' and $\text{Mn\_ES}(50).05$ increases this probability and higher PctCapitellidae lowers it. By subtracting this from eq. (1) we get an index where low values indicate low BEQ. The APBI was then scaled to the range 0–10 by multiplying by 10:

$$\text{APBI} = 10 \times (1 - p)$$

This index was developed to encompass a wide range of habitats and conditions; nevertheless, according to the authors, the choice of a smaller subset of data (e.g. mud or a smaller geographic area) will lower the variability and result in a more accurate indicator. The APBI has been applied in the scope of the NCA Northeast report (USEPA, 2006) and National Coastal Condition Report III (USEPA, 2007); nevertheless, it should undergo a series of validations and calibrations processes to be accepted as a universal index (Hale & Heltshe, 2008). The authors also point out that the efficiency of this index is unknown for low-salinity areas, and since it was designed to be applied in soft-bottom communities it has a higher discriminating impact in mud than in sand areas. Furthermore, this index has been developed using summer data, and so the seasonality effects should be assessed as well. Nonetheless, this index also pretended to examine if the Signal Detection Theory can help to evaluate the ability of the APBI to detect a degraded benthic environment, demonstrating that it can be used as a guide in the decisions that environmental managers have to take about thresholds and where to assign resources (Hale & Heltshe, 2008). In addition, the PPV–NPV techniques can be used to foresee how well an index developed for one geographic area might work in another region with different incidence of degraded conditions.

2.3.5.2 Benthic Condition Index (Engle *et al.*, 1994; Engle & Summers, 1999)

The Benthic Condition Index was designed to evaluate the environmental condition of degraded systems comparatively to reference situations (non-degraded conditions) based on the response of benthic organisms to environmental stressors. This index, which results from the refinement of a previous attempt (Engle *et al.*, 1994), reflects the benthic-community responses to perturbations in the natural system (Engle, 2000). The benthic index includes: (1) Shannon–Wiener Diversity Index adjusted to salinity; (2) mean abundance for Tubificidae; (3) percentages of abundance of the class Bivalvia; (4) percentages of abundance of the family Capitellidae and (5) percentages of abundance of the order Amphipoda.

To calculate this index, one first needs to calculate the expected Shannon–Wiener diversity index, according to the bottom salinity:

$$H'_{expected} = 2.618426 - (0.044795 \times salinity) + (0.007278 \times salinity^2)$$
$$+ (-0.000119 \times salinity^3)$$

The final Shannon–Wiener's score is given by dividing the observed by the expected diversity values. After the calculation of the abundance and proportions of the organisms involved, it is necessary to log transform the abundances and arcsine transform the proportions. Based on this, the discriminant score is calculated as:

$$\begin{aligned}
\text{Discriminant score} = \ &(1.5710 \times \text{Proportion of expected diversity}) \\
&+ (-1.0335 \times \text{Mean abundance of Tubificidae}) \\
&+ (-0.5607 \times \text{Percent Capitellidae}) \\
&+ (-0.4470 \times \text{Percent Bivalvia}) \\
&+ (0.5023 \times \text{Percent Amphipoda})
\end{aligned}$$

To turn the index in a practicable and easily understood measure by policy-makers, the final benthic index score is given by:

$$\text{Benthic index} = \left(\frac{\text{Discriminant score} - (-3.21)}{7.50} \right) \times 10$$

where -3.21 is the minimum of the discriminant score, and 7.50 is the range of the discriminant score.

When a community is affected by contaminants, the benthic organisms diminish in abundance and number of species, while there is an increase in the abundance of opportunistic or pollution-tolerant species. After the discriminant score transformation, the benthic index can range between 0 and 10. Therefore, values lower than 3 indicate degraded conditions, values higher than 5 indicate non-degraded sites and values between 3 and 5 reveal transition conditions. According to the authors, this index classifies benthic communities' condition within and among estuaries.

2.3.5.3   BHQ (Nilsson & Rosenberg, 1997)

The BHQ was designed to evaluate the environmental condition of the soft-bottom habitat quality of Havstensfjord (Baltic Sea) through analysis of sediment profile and surface images (SPI). The BHQ Index relies on the relation between the classical distribution of benthic infaunal communities and the organic enrichment, based on the Pearson and Rosenberg model (1978). This tool integrates the structures on the sediment surface, structures in the sediment and the redox potential discontinuity images (RPD). Therefore, the parameterisation of sediment and animal features may be a useful combination to describe and assess habitat quality (Rhoads & Germano, 1986). This index attempts to show the usefulness of SPI in demonstrating benthic habitat changes connected with physical disturbance, specifically with low oxygen concentrations (Nilsson & Rosenberg, 2000). The calculation of the BHQ Index from the sediment profile images can be computed by:

$$BHQ = \sum A + \sum B + C$$

where A is the measure of surface structures, B the measure of the subsurface structures and C the mean sediment depth of the apparent RPD. The parameters used in this index were all measured from the images and the scoring could be seen as an objective assessment of the successional stages. Deep subsurface activity, such as feeding voids and many burrows, which often is associated with a thick RPD, have a high scoring and contribute to a high BHQ Index. It can range between 0 and 15, where high scores are associated with mature benthic faunal successional stages and low scores with pioneering stages or azoic bottoms. According to Rosenberg et al. (2004), the BHQ Index could also be a useful tool for the WFD implementation in assessing the BHQ. Therefore, instead of the earlier separation of the BHQ index into four successional stages, Rosenberg et al. (2004) underpin the division into five classes in accordance with the WFD requirements. According to the index authors, this scoring method can be valid for many boreal and temperate areas, as in these areas the benthic infauna is similarly structured and has a similar distribution and activity within the sediment (Pearson & Rosenberg, 1978; Rhoads & Germano, 1986). Despite sharing some principles, one main difference that distinguishes these two indices is that in the OSI (Rhoads & Germano, 1986; see section 2.3.5.14) the successional stages are determined by examining the images by eye whereas in the BHQ Index the different structures in the images are scored and their summary relates to a particular community stage (Nilsson & Rosenberg, 2000).

One advantage enumerated by the BHQ authors is the fact that a benthic quality assessment based on numerical scoring, as the one used in the BHQ Index, allows statistical comparisons between strata and communities. Furthermore, this method can also roughly forecast oxygen regimes over integrative time scales, becoming a useful tool for environmental managers interested in benthic assessment and in rough but quantitative approximation of near-bottom Dissolved Oxygen regimes (Cicchetti *et al.*, 2006). Moreover, the use of SPI methods is a rapid and inexpensive way of tracking and assessing the BHQ and is very useful in characterizing the successional stages of the organic enrichment gradient (Nilsson & Rosenberg, 2000; Wildish *et al.*, 2003). Nevertheless, according to Wildish *et al.* (2003) there are some benthic habitats where this method cannot be applied, as for example in areas where soft bottoms are absent and where coarse sediments or rock predominate; or even in areas where water depth exceeds reasonable SCUBA diving depths (approximately 30 m).

2.3.5.4   Benthic Index of Biotic Integrity (B-IBI) (Weisberg *et al.*, 1997)

The Chesapeake Bay B-IBI (Weisberg *et al.*, 1997) integrates several benthic attributes, related to healthy benthic-community structure, to calculate the global condition of a region. It gives the actual status of the benthic community as a function of its deviation from the reference condition. Therefore it can provide trends within the system once calibrated to reference conditions. The indicators used to calculate the benthic index are: (1) Shannon–Wiener species-diversity index; (2) total species abundance; (3) total species biomass; (4) percent abundance of pollution-indicative taxa; (5) percent abundance of pollution-sensitive taxa; (6) percent biomass of pollution-indicative taxa; (7) percent biomass of pollution-sensitive taxa; (8) percent abundance of carnivore and omnivore species; (9) percent abundance of deep-deposit feeders; (10) Tolerance Score and (11) Tanypodinae to Chironomidae percent abundance ratio.

To calculate the index metrics, several steps have to be implemented (Llansó *et al.*, 2002b). The epifaunal species as well as other individuals that are not representative of the subtidal communities have to be eliminated from the species lists and from the calculation (e.g. Nematoda or fish species). To calculate the diversity measure (H'), higher taxonomic groups have to be retained as well (e.g. Polychaeta or Amphipoda). The sensitive/indicative pollution species can be classified according to the *AMBI* species classification, considering the I and II ecological groups as sensitive to pollution and the ecological groups IV and V as pollution-indicative species. When field and experimental data are not available for feeding strategies metrics (% of carnivore–omnivore species and % subsurface deposit feeders), a literature review can be made to classify all the species.

Although the B-IBI Index integrates 11 metrics, not all of them are used here to calculate the overall ecosystem score and condition. Llansó *et al.* (2002a) defined seven major estuarine stretches for the Chesapeake Bay, according to the Venice transitional water organization scheme for salinity and sediment types. Depending on the estuarine stretch under analysis, different metrics are used to estimate the local condition and status (Table 11).

Table 11: Thresholds used to score each B-IBI metric.

	Scoring criteria		
	5	3	1
Tidal Freshwater			
Shannon–Wiener	≥1.8	1–1.8	<1
Abundance (g m^{-2})	≥1000–4000	500–1000 or ≥4000–10,000	<500 or ≥10,000
Biomass (g m^{-2})	≥0.5–3	0.25–0.5 or ≥3–50	<0.25 or ≥50
Abundance pollution-indicative taxa (%)	≤25	25–75	>75
Oligohaline			
Shannon–Wiener	≥2.5	1.9–2.5	<1.9
Abundance (m^{-2})	≥1500–3000	500–1500 or ≥3000–8000	<500 or ≥8000
Biomass (g m^{-2})	≥3–25	0.5–3 or ≥25–60	<0.5 or ≥60
Abundance pollution-indicative taxa (%)	≤25	25–75	>75
Abundance sensitive taxa (%)	≥40	10–40	<10
Low Mesohaline			
Shannon–Wiener	≥2.5	1.7–2.5	<1.7
Abundance (m^{-2})	≥1500–2500	500–1500 or ≥2500–6000	<500 or ≥6000
Biomass (g m^{-2})	≥5–10	1–5 or ≥10–30	<1 or ≥30
Abundance pollution-indicative taxa (%)	≤10	10–20	>20
Biomass pollution-sensitive taxa (%)	>80	40–80	<40
Biomass >5 cm below sediment–water interface (%)	≥80	10–80	<10
High Mesohaline Sand			
Shannon–Wiener	≥3.2	2.5–3.2	<2.5
Abundance (m^{-2})	≥1500–3000	1000–1500 or ≥3000–5000	<1000 or ≥5000
Biomass (g m^{-2})	≥3–15	1–3 or ≥15–50	<1 or 50

Abundance pollution-indicative taxa (%)	<10	10–25	>25
Abundance sensitive taxa (%)	≥40	10–40	<10
Abundance carnivores & omnivores (%)	≥35	20–35	<20
High Mesohaline Mud			
Shannon–Wiener	≥3	2–3	<2
Abundance (m⁻²)	≥1500–3000	1000–1500 or ≥2500–5000	<1000 or ≥5000
Biomass (g m⁻²)	≥2–10	0.5–2 or ≥10–50	<0.5 or ≥50
Biomass pollution-indicative taxa (%)	≤5	5–30	>30
Biomass pollution-sensitive taxa (%)	≥60	30–60	<30
Abundance carnivores & omnivores (%)	<25	10–25	<10
Biomass >5 cm below sediment–water interface (%)	≥60	10–60	<10
Polyhaline Sand			
Shannon–Wiener	≥3–5	2.7–3.5	<2.7
Abundance (m⁻²)	≥3000–5000	1500–3000 or ≥5000–8000	<1500 or ≥8000
Biomass (g m⁻²)	≥5–20	1–5 or ≥20–50	<1 or ≥50
Abundance pollution-indicative taxa (%)	≤5	5–15	>15
Biomass pollution-sensitive taxa (%)	≥50	25–50	<25
Abundance deep-deposit feeders (%)	>25	10–25	<25
Polyhaline Mud			
Shannon–Wiener	>3.3	2.4–3.3	<2.4
Abundance (m⁻²)	≥1500–3000	1000–1500 or ≥3000–8000	<1000 or ≥8000
Biomass (g m⁻²)	≥3–10	0.5–3 or ≥10–30	<0.5 or ≥30
Biomass pollution-indicative taxa (%)	≤5	5–20	>20
Biomass pollution-sensitive taxa (%)	≥60	30–60	<30
Abundance carnivores & omnivores (%)	≥40	25–40	<25
Taxa > 5 cm below sediment–water interface (%)	≥40	10–40	<10

Moreover, the tolerance score and the percentage of Tanipodinae to Chirono-midae metrics were considered by the index authors as facultative, as long as no score was attributed to them and that, for the oligohaline zone, the lack of the two metrics was considered in the index average calculation (removed from the denominator factor). This index classification gives scores for the different indicators in relation to the reference conditions. When the two situations are identical, a score of 5 is given and when they are very different a score of 1 is attributed. The final index score is determined by the average of the individual scores (Llansó et al., 2002b). Values inferior to 2 correspond to severely degraded conditions, between 2.1 and 2.6 to degraded conditions, ranging from 2.7 and 2.9 to marginal condition and, finally, values higher than 3 indicate that the area meets restoration goals.

Other indices have been developed having the B-IBI as a role model, never-theless adapted for other geographical systems. For instance, Van Dolah et al. (1999) modified this index (Carolina Province B-IBI) and it is calculated using the average score of the following metrics: mean abundance; mean number of taxa; percentage of abundance of the top two numerical dominants and percent-age abundance of pollution-sensitive taxa.

2.3.5.5   Benthic Quality Index (BQI) (Rosenberg et al., 2004)

The BQI was designed to assess environmental quality according to the WFD. Tolerance scores, abundance and species-diversity factors are used in its deter-mination. The main objective of this index is to attribute tolerance scores to the benthic fauna to determine their sensitivity to disturbance. The index is expressed as:

$$BQI = \left\{ \sum \left(\frac{A_i}{TotA} \right) \times ES50_{0.05i} \right\} \times {}^{10}Log(S + 1)$$

where $A_i/Tot\,A$ is the mean relative abundance of this species, and $ES50_{0.05i}$ the tolerance value of each species, i, found at the station. This metric corresponds to 5% of the total abundance of this species within the studied area. Further, the sum is multiplied by base 10 logarithm for the mean number of species (S) at the station, as high species diversity is related to high environmental quality. The goal of using the values calculated from the 5% lowest abundance of a particular species ($ES50_{0.05}$) is that this value is assumed to be representative for the great-est tolerance level for that species along an increasing gradient of disturbance, that is if the stress slightly increases that species will disappear. This method is similar to that proposed by Gray & Person (1982) and presents the advantage of reducing the weight of outliers during the index calculation. This parameter can be computed as:

$$ES50 = 1 - \sum_{i=1}^{s} \frac{(N - N_i)!(N - 50)!}{(N - N_i - 50)!N!}$$

where N is the total abundance of individuals, N_i the abundance of the ith species and s the number of species at the station.

Results from this analysis can range between 0 and 20 (reference value) according to the classification made by the WFD for the coastal environmental status (1 to <4: Bad; 4 to <8: Poor; 8 to <12: Moderate; 12 to <16: Good; 16 to <20: High).

Two methodological constraints of this index can be highlighted: the sample area is not the same among sampling protocols, and individuals' distribution among species may not be random, particularly when some species appear as strong dominants. Thus, Rosenberg *et al.* (2004) recommend the use of many stations and replicates for the quality assessment of an area. Moreover, according to Zettler *et al.* (2007) this index presents strong correlations with environmental variables, such as salinity, decreasing the scores with decreasing salinities. Furthermore, this index requires regional datasets (the $ES50_{0.05}$ calculation is based on the specific framework of species present in the study area), and the delimitation of local reference values, depending on the areas under study such that different maximum values can be achieved (Rosenberg *et al.*, 2004; Reiss & Kröncke, 2005; Labrune *et al.*, 2006; Zettler *et al.*, 2007).

2.3.5.6   BRI (Smith *et al.*, 2001)

The BRI was developed for the Southern California coastal shelf and is a marine analogue of the Hilsenhoff Index used in freshwater benthic assessments (Hilsenhoff, 1987). This index is calculated using a two-step method in which ordination analysis is employed to establish a pollution gradient. The pollution tolerance of each species is later determined based upon its abundance along the gradient (Smith *et al.*, 1998). The index main goal is to establish the abundance-weighted average pollution tolerance of the species in a sample, which is considered a very useful screening tool (Bergen *et al.*, 2000). The basis of this index is that each species has a tolerance for pollution and if that tolerance is known for a large set of species, then it is possible to infer the degree of degradation from species composition and its tolerances (Gibson *et al.*, 2000). The index can be given as:

$$I_s = \frac{\sum_{i=1}^{n} p_i \sqrt[3]{a_{si}}}{\sum_{i=1}^{n} \sqrt[3]{a_{si}^f}} \tag{a}$$

where I_s is the index value for the sample s, n the number of species in the sample s, p_i the tolerance value for species i (position on the gradient of pollution) and a_{si} the abundance of species i in sample s. The exponent f is for transforming the abundance weights. So, if $f = 1$, the raw abundance values are used; if $f = 0.5$, the square root of the abundances are used and if $f = 0$, I_s is the arithmetic value of the p_i values greater than zero, once that all $a_{si}^f = 1$ (Smith *et al.*, 2001). The average position for each species (p_i) on the pollution gradient defined in the ordination space is measured as:

$$p_i = \frac{\sum_{j=1}^{ti} g_{ij}}{ti} \tag{b}$$

where ti is the number of samples to be used in the sum, with only the highest ti species abundance values included in the sum. The g_{ij} is the position of the species i on the ordination gradient for sample j. The p_i values obtained in eq (b) are used as pollution-tolerance scores in eq (a) to compute the index values.

This index provides a quantitative scale ranging from 0 to 100, where low scores are indicative of healthier benthic communities (i.e., community composition most similar to that occurring at not-impacted regional reference sites). The BRI scoring defines four levels of response beyond reference condition (0–33: marginal deviation; 34–43: loss of biodiversity; 44–72: loss of community function; >72: defaunation).

Although it can be useful to quantify disturbances, it is not able to distinguish between natural and anthropogenic disturbance, such as the natural impacts that river flows may have on benthic communities (Bergen *et al.*, 2000). Nevertheless, this index presents the advantage of not underestimating biological effects, as well as presenting low seasonal variability (Smith *et al.*, 2001).

2.3.5.7   IBI for fishes (McGinty & Linder, 1997)

A fish-based IBI was developed for tidal fish communities of several small tributaries of the Chesapeake Bay (Jordan *et al.*, 1990; Vaas & Jordan, 1990; Carmichael *et al.*, 1992).

Nine metrics are used to calculate the index bearing in mind species richness, trophic structure and abundance: number of species; number of species comprising 90% of the catch; number of species in the bottom trawl; proportion of carnivores; proportion of planktivores; proportion of benthivores; number of estuarine fish; number of anadromous fish and total fish with Atlantic menhaden removed.

The quantification of the different metrics utilized to estimate the index is carried out by comparing the value of a metric from the sample of unknown water quality to thresholds established from reference data distributions.

2.3.5.8   Coefficient of pollution (Satsmadjis, 1982, 1985)

This index, proposed in 1982 by J. Satsmadjis, recognizes the relationship between the infaunal and sediment structure. The calculation involves the following assumptions: (1) the number of species increases linearly with the number of individual animals; (2) sediment structure can be represented by a single value (the sand equivalent, s') based on the relative percentages of sand and silt and (3) faunal abundance is related to s' and depth. The following integrated equations are used:

$$S' = \frac{s+t}{(5+0,2s)}$$

$$i0 = (-0,0187s'2 + 2,63s' - 4)(2,20 - 0,0166\,h)$$

$$g' = \frac{i}{(0,0124i + 1,63)}$$

$$P = \frac{g'}{[g(i/i0)1/2]}$$

where: P = coefficient of pollution; S' = sand equivalent; S = percentage of sand; t = percentage of silt; i_0 = theoretical number of individuals; i = number of individuals; h = station depth; g' = theoretical number of species and g = number of species. Coefficients of 1.5–2.0, 2.0–3.0, 3.0–4.0 and 4.0–8.0 are assumed to indicate Slight, Moderate, Heavy and Very Heavy pollution, respectively.

2.3.5.9   EQI (Fano et al., 2003)

This index is based on the characteristics of the primary producers and of the benthic faunal communities and has been designed to assess the ecological quality of coastal lagoons. The following parameters are taken into account: macrofaunal abundance; number of taxa; taxonomic diversity; functional diversity; macrofaunal biomass; phytoplankton biomass and macrophytal biomass. Each one of these attributes, which are expressed by heterogeneous units, is then transformed into a dimensionless quality scale ranging from 0 to 100, simply by assigning 100 to the highest value, and by normalizing to 100 all the other values. Once all attributes are expressed by means of this scale, they are combined to obtain the integrated index, whose maximum theoretical value will vary from 700 to 800, depending on whether macroalgae are present in a particular habitat. These values would correspond to the optimum condition of the index, irrespective of the units and magnitudes used to measure the different individual attributes. Obviously, the closer the actual values are to, say, 800, the better the condition of the environment.

EQI also allows for comparisons between sites from different lagoons (nEQI). Data sets from the different lagoons are merged into a worksheet so that the value of each attribute can be rescaled, using the same quality scale of 0–100 on the complete data set. Finally, scores are summed and divided by the number of attributes measured in each different lagoon. In this way, the use of EQI can derive a series of continuous values, from 0 to 800 (nEQI: from 0 to 100). The result obtained is a functional classification of the sites within a lagoon or between different lagoons.

Up to now, this index has only been applied in three coastal lagoons in Italy (Sacca di Goro, Valle Fattibello and Valli di Comacchio).

2.3.5.10   EBI (Deegan et al., 1993)

The EBI includes the following eight metrics: total number of species; dominance; fish abundance; number of nurseries; number of estuarine spawning species; number of resident species; proportion of benthic associated species and proportion of abnormal or diseased fish.

The usefulness of this index requires that it reflect not only the current status of fish communities but also its applicability over a wide range of estuaries, although this is not entirely achieved (Bettencourt et al., 2004).

2.3.5.11   FHI (Cooper et al., 1993)

This index is based on the Community Degradation Index (CDI) developed by Ramm (1988, 1990) which measures the degree of dissimilarity (degradation) between a potential fish assemblage and the real measured fish assemblage.

The FHI provides a measure of similarity (health) between the potential and actual fish assemblages and is calculated using the formula:

$$FHI = 10(J)\left[\frac{Ln(P)}{Ln(P_{max})}\right]$$

where J = the number of species in the system divided by the number of species in the reference community; P = the potential species richness (number of species) of each reference community and P_{max} = the maximum potential species richness from all the reference communities. The index ranges from 0 (Poor) to 10 (Good).

The FHI was used to assess the state of South African estuaries (Cooper *et al.*, 1993; Harrison *et al.*, 1994, 1995, 1997, 1999). Although the index has proved to be a useful tool in condensing information on estuarine fish assemblages into a single numerical value, the index is based only on presence/absence data, and consequently does not take into account the relative proportions of the various species present.

2.3.5.12 M-AMBI (Muxika *et al.*, 2007)

This integrative formula was designed in response to the WFD requirements to include several metrics describing the benthic-community integrity (e.g. abundance, biomass or diversity measures) and parameters that are considered to define better the water bodies' EcoQS. Moreover, it is intended to support the European Marine Strategy Directive (Borja, 2006), in assessing the ecological status of continental shelf and oceanic water bodies (Muxika *et al.*, 2007). The M-AMBI is a combination of the proportion of 'disturbance-sensitive taxa', through the computation of the *AMBI* Index, species richness (it uses the total number of species, *S*) and diversity through the use of the Shannon–Wiener Index, which overcame the need to use more than one index to evaluate the overall state and quality of an area (Zettler *et al.*, 2007). These parameters are integrated through the use of Factorial Analysis (FA) and Discriminant Analysis (DA) techniques. This method compares monitoring results with reference conditions by salinity stretch, for estuarine systems, to derive an Ecological Quality Ratio (EQR). These final values express the relationship between the observed values and reference condition values. At 'high' status, the reference condition may be regarded as an optimum where the EQR approaches the value of one. At 'Bad' status, the EQR approaches the zero value. The M-AMBI analysis relies on the Euclidean Distance ratio between each area and the reference spots, together with the distance between High and Bad status reference condition (Muxika *et al.*, 2007). The stations are located between the reference conditions and have M-AMBI values ranging from 0 to 1. The boundaries that allow the distinction of the five ecological states are the following: >0.82 High; 0.62–0.82 Good; 0.41–0.61 Fair; 0.20–0.40 Poor; <0.20 Bad.

The M-AMBI has been the outcome of the Intercalibration process among states members for the WFD common methodologies; nevertheless, it has been applied to other systems outside Europe, such as in Chesapeake Bay, United States of America, where it revealed to be a consistent measure, providing high-agreement percentages with local indices (Borja *et al.*, 2008). A main advantage attributed to this index, as well as of *AMBI*, is that both are easily computed, and the software can be freely downloaded at http://www.azti.es. Moreover, the M-AMBI seems to provide a more accurate system classification in low-salinity habitats, than the *AMBI* alone (Muxika *et al.*, 2007; Borja *et al.*, 2008).

2.3.5.13   OSI (Rhoads & Germano, 1986)

The OSI was developed to assess the BHQ in shallow water environments, allowing the evaluation of stages of organic pollution in the ecosystem. This index presents four main metrics: (1) Dissolved Oxygen conditions; (2) depth of the apparent RPD; (3) infaunal successional stage and (4) presence or absence of sedimentary methane. The successional stage was measured with sediment profile images, which characterises the benthic habitat in relation to physical–chemical features (Rhoads & Germano, 1982). The OSI Index has also been used in some studies to map habitat quality (Rhoads & Germano, 1986), to assess physical disturbances and organic enrichment (Valente *et al.*, 1992) and to evaluate the effects of mariculture (O'Connor *et al.*, 1989). Two recent studies showed that low values of apparent RPD were correlated with low OSI scores (Nilsson & Rosenberg, 2000).

Rhoads & Germano (1986) based their index on the mean depth of the apparent RPD, the presence of gas voids and on a visual classification of the infauna into successional stages, which could range from -10 to $+11$. The lowest values are attributed to the bottom sediments with low or no Dissolved Oxygen, without apparent macrofauna, and with methane present in the sediment. The highest values are attributed to aerobic sediments with a deep apparent RDP, established macrofaunal communities and without methane gas. The index classification is as follows: <0: degraded benthic habitat; 0 to <7: disturbed benthic habitat; 7–11: undisturbed benthic habitat.

2.3.5.14   PLI (Jeffrey *et al.*, 1985)

This index, based on contamination loads, includes three parameters: (1) water; (2) fauna and (3) flora and sediment. The PLI scores individual sediment contaminants according to a log scale from baseline to threshold (Wilson, 2003). The formula is computed as:

$$PLI = \text{anti}\log_{10}\left\{1 - \left(\frac{CP - B}{T - B}\right)\right\}$$

where CP is the pollutant concentration; B the unpolluted baseline and T the damage threshold. The scores for each pollutant are summed to give a total site PLI thus:

$$\text{Site PLI} = (PLI_1 \times PLI_2 \times PLI_n)^{1/n}, \quad \text{for } n \text{ pollutants}$$

The sites scores are then summed likewise to give the estuary index value:

$$PLI = (PLI_1 \times PLI_2 \times PLI_j)^{1/j}, \quad \text{for } j \text{ sites}$$

The PLI varies from 0 (highly polluted) to 10 (unpolluted). This index allows the comparison between several estuarine systems and has been applied in several geographical regions, such as in Europe (Wilson *et al.*, 1987) and US estuaries (Wilson, 2003). Caeiro *et al.* (2005) have highlighted the ease of implementation of this index.

2.3.5.15 P-BAT (Teixeira *et al.*, submitted)

The P-BAT integrates, in a cumulative index, three widely used metrics – Shannon–Wiener Index (H'), Margalef Index (d) and *AMBI* – which are based on different approaches when evaluating system status. This integration results from works on the Portuguese transitional and coastal waters systems that demonstrate that, when evaluating the system condition, the combination of several metrics is more accurate than single metrics. The Shannon–Wiener Index is a diversity measure that takes into account the proportional abundance of species; the Margalef Index is based on the specific richness of a system and the *AMBI* on the ecological strategies followed by estuarine organisms (indicator species). Overall index classifications have been developed for the Portuguese coastal and transitional water bodies (soft bottom). The values can range between 0 (Bad ecological quality) and 1 (Good ecological quality).

To calculate the multimetric approach, the Shannon–Wiener, Margalef and *AMBI* values (previously calculated) were standardised by subtracting the mean and dividing by the standard deviation. Afterwards, a FA was conducted to construct a three coordinate system that was then used to derive the final station score, using as comparison the reference conditions determined for the system. These reference conditions were estimated based on two opposite situations: the best condition that a system could present (without impacts) versus the worst-possible scenario for the same system (Bald *et al.*, 2005; Muxika *et al.*, 2007). In the analysis, the Varimax rotation method is adopted to make it easier to interpret the results. From this analysis, the scores of the first three factors are extracted (using the 'factor loadings' on the calculation). After obtaining the sampling stations' relative position (scores extracted from the FA), the projection of each sampling station in the axis connecting both reference stations (High and Bad status) is calculated in the new three-dimensional space created by the FA (Figure 3).

Subsequently, the Euclidean Distance of each projection to the virtual station corresponding to Bad status is measured. The value of 1 (accordingly to the definition of EQR in the WFD) is attributed to the distance between both virtual reference stations (Bad and High). As a result, stations in better condition, with higher Ecological Status, will achieve values near 1, while stations in a worse ecological condition will be located nearer the bad reference station and will assume values closer to 0.

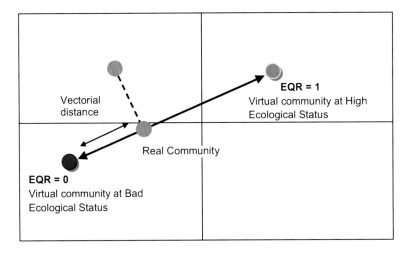

Figure 3: Principal Components Analysis showing virtual communities at High and Bad Ecological Status, and relative position of the real community for derivation of the EQRs (after Bald *et al.*, 2005).

The boundaries along this EQR axis (from 0 to 1) were defined, reflecting the five EcoQs classes stated in WFD: Bad, Poor, Moderate, Good and High. After the Intercalibration exercise, performed among European Member States for coastal waters, the following thresholds to define EQS classes were proposed (Teixeira *et al.*, submitted): 0–0.27 Bad; 0.28–0.44 Poor; 0.45–0.58 Moderate; 0.59–0.79 Good and 0.80–1 High EQS. Boundaries for transitional waters were already proposed but similar to what was done for coastal waters (Borja *et al.*, 2007), these thresholds will be agreed upon after the Intercalibration exercise between EU Member States.

2.3.5.16   Portuguese marine macroalgae assessment tool (P-MarMAT) (Neto *et al.*, submitted)

The P-MarMAT is a multimetric approach based in seven metrics: macroalgae species richness, proportion of Chlorophyceae (greens), number of Rhodo-phyceae (reds), Ecological Status Groups (ESG) ratio (Orfanidis *et al.*, 2001, 2003; Wells *et al.*, 2007), proportion of opportunists, coverage of opportunists and shore description. The first step for implementing the methodology is to elaborate the Selected Species List (SSL), representative of local macroalgal natural occurrences, which will be the basic of application of metrics. Each of the selected metrics varies over a range of values that depends on their own characteristics. The metric's range is divided into five quality classes that, depending on the respective boundary and following the Normative Defini-tions (Annex V in WFD, 2000/60/CE), score successively from zero (Bad) to four (High). The score achieved by each of individual metric is then integrated in a 'Sum of Scores'. Each metric may attain different levels in the quality scale, where Bad scores 0, Poor scores 1, Moderate scores 2, Good scores 3 and High scores 4. The 'species richness' and 'coverage of opportunists' are double weighted, counting twice to the sum of scores. As an example, if 10 species are found in a shore, it scores 2 for species richness; if the proportion of greens is 15, it scores 3 and so on for the other metrics. The sum of all these metrics' scores constitutes the 'Sum of Scores' (Table 12).

The EQR is calculated as:

$$EQR = \frac{Sum\,of\,Scores}{36}$$

converting the 'Sum of Scores' in a 0 to 1 scale, accordingly to the definition of 'EQR' in the WFD (2000/60/CE).

Shores in better condition, meaning higher Ecological Status, will achieve values near 1, while the ones in worse ecological condition will be located nearer bad reference station and will assume values nearer 0. The 'EQR' is later translated into 'EQS' classes (Bad, Poor, Moderate, Good and High) by the boundaries showed in Table 12. In this approach, equidistant intervals (0.2) were used for 'EQR' boundaries. Probably some future adjustment will be required to improve the final result on assessing intertidal rocky sea shores.

Table 12: Boundaries' criteria considered for each selected metric, sum of scores and EQR.

Metrics	Bad	Poor	Moderate	Good	High
Species richness[a]	0–5	5–8	9–16	17–24	≥ 25
Proportion of Chlorophyceae	40–100	30–40	20–30	10–20	0–10
Number of Rhodophyceae	0–5	6–10	11–15	16–20	≥ 21
ESG Ratio	0–1	1–1.5	1.5–2	2–2.5	> 2.5
Proportion of Opportunists	40–100	30–40	20–30	10–20	0–10
Coverage of opportunists (%)[a]	70–100	30–70	20–30	10–20	0–10
Shore Description	-	15–18	12–14	8–11	1–7
Sum of Scores	0–7	8–14	15–21	22–28	29–36
EQR	0–0.2	0.2–0.4	0.4–0.6	0.6–0.8	0.8–1
EQS	Bad	Poor	Moderate	Good	High

Note: For the 'Metrics' part, Bad scores 0, Poor scores 1, Moderate scores 2, Good scores 3, and High scores 4, to calculate de 'Sum of Scores'. Translation of the achieved EQR in the EQS (Bad, Poor, Moderate, Good or High) when assessing rocky shores' ecological quality.

[a] Factor of 2, counts twice in the metrics sum of scores calculation.

2.3.5.17 TRIX (Vollenweider *et al.*, 1998)

This index is based on chlorophyll *a*, oxygen saturation, mineral nitrogen and total phosphorus. To be selected and included in this trophic area, the parameters have to be directly related to eutrophication phenomena and the index aims at characterising the trophic state of coastal waters. It is expressed by the following equation:

$$\text{TRIX} = \frac{k}{n} \times \sum \frac{(M_i - L_i)}{(U_i - L_i)}$$

in which $k = 10$ (scaling the result between 0 and 10), $n = 4$ (number of variables integrated, M_i = measured value of variable i, U_i = upper limit of variable i, L_i = lower limit of value i.

The resulting TRIX values are dependent on the upper and the lower limits chosen and indicate how close the current state of a system is to the natural state. However, the comparison of TRIX values obtained for different areas becomes more difficult. In general, when a wide, more general range is used for the limits, TRIX values for different areas are more easily compared to each other. According to Giovanardi & Vollenweider (2004) and Penna *et al.* (2004), values ranging from 0 to 4 correspond to high-quality, 4–5 to good, 5–6 to moderate and 6–10 to degraded conditions.

2.3.5.18 VPBI (Paul *et al.*, 2001)

This VPBI has been developed over two stages for application in the Virginia province, United States of America. Here, we detail the most-recent formulation by Paul *et al.* (2001) which is an expanded version of Schimmel *et al.* (1999). The goal of VPBI is to evaluate the benthic condition of estuarine communities, discriminating between degraded and non-degraded sites. This index is based on a measure of diversity (related with not-impacted sites) and the abundance of pollution-tolerant taxa, Tubificidae and Spionidae (related with impacted conditions). The index was developed for the US EPA Environmental Monitoring and Assessment Program. The three benthic metrics in the index are: (1) salinity-normalised Gleason's, D, based upon infauna and epifauna; (2) salinity-normalised Tubificidae abundance and (3) abundance of Spionidae. This index is given by:

Benthic Index = 1.389 × (salinity-normalised Gleason's, D,
based upon infauna and epifauna − 51.5)/28.4 − 0.651
× (salinity-normalised Tubificidae abundance − 28.2)/119
− 0.375 × (Spionidae abundance − 20.0)/45.4 = 0.0489
× salinity-normalised Gleason's, D − 0.00545
× salinity-normalised Tubificidae abundance
− Spionidae abundance − 2.20

where salinity-normalised Gleason's, D, based upon infauna and epifauna = Gleason's $D/(4.283 - 0.498 \times$ bottom salinity $+ 0.0542 \times$ bottom salinity$^2 - 0.00103 \times$ bottom salinity$^3) \times 100$ and salinity-normalised Tubificidae abundance

= Tubificidae abundance $- 500 \times \exp (-15 \times$ bottom salinity), and $\exp (...)$ denotes the exponential function.

The salinity normalization for Tubificidae abundance required a different procedure than that used to normalize the other benthic metrics. Tubificidae are only observed for low-salinity water with some occurrence being normal for not-impacted sites. Impacted sites would be characterised by abundance of Tubificidae. This index identifies the not-impacted sites by a negative gradient of salinity-normalised abundance and the impacted sites with positive values. In this index, positive values are indicative of healthy community conditions and negative values reflect degraded communities.

Although this index gives an overall system condition, it is important to notice that it was based on the benthic communities, so the habitat condition of the pelagic area, submerged aquatic vegetation and marshes is not assessed.

2.3.6   Indices thermodynamically oriented or based on network analysis

In the last two decades, several functions have been proposed as holistic ecological indicators, intending (1) to express emergent properties of ecosystems arising from self-organisation processes in the run of their development, and (2) to act as orientors (goal functions) in the development of models, as mentioned above. Such orientors intend to account for suitable system-oriented characteristics, expressing natural tendencies of an ecosystem's development (Marques *et al.*, 1998).

In general, these proposals resulted from a wider application of theoretical concepts, following the assumption that it is possible to develop a theoretical framework able to explain ecological observations, rules and correlations based on an accepted pattern of ecosystem theories (Jørgensen & Marques, 2001). For instance, the case of ascendency (Ulanowicz, 1980, 1986; Ulanowicz & Norden, 1990), and Exergy (Jørgensen & Mejer, 1979, 1981), a concept derived from the field of thermodynamics, which can be seen as energy with a built-in measure of quality, which has been tested in several studies (e.g. Nielsen, 1990; Jørgensen, 1994; Fuliu, 1997; Marques *et al.*, 1997, 2003).

2.3.6.1   Ascendency (Ulanowicz, 1980)

The emphasis on ecology has been shifting towards a vision of ecosystems as a system of interactions (Fasham, 1984; Frontier & Pichod-Viale, 1995), meaning that the centre of interest has been diverted from the state of the biomass of the different groups of organisms, and the focus is now on the status of the interactions between them, as quantified by flows of matter or energy (Niquil *et al.*, 1999).

Any index used in such attempts must combine the attributes of a system's activity level and community structure. One such measure derives from the analysis of networks of trophic exchanges and is called the 'ascendency' system. Ulanowicz (1980) defines ascendency as an index that quantifies both the level of system activity and the degree of its organization, whereby it processes material in an autocatalytic fashion.

Ascendency is a rather abstract concept that nevertheless reveals manifold attributes when viewed from a variety of aspects. Ascendency was originally created to quantify the developmental status of an ecosystem. If one suspects that a particular disturbance has negatively impacted an ecosystem, ascendency can be invoked to test that hypothesis quantitatively, provided sufficient data are available to construct networks of exchanges before and after the impact. Not only can one make before and after comparisons, but the developmental stages of disparate ecosystems can also be compared with one another (e.g. Ulanowicz & Wulff, 1991).

Using ascendency, makes it possible to determine quantitatively whether a system has grown or shrunk, developed or regressed. Furthermore, particular patterns of changes in the information variables can be used to identify processes that hitherto had been described only verbally (Ulanowicz, 2000). For example, eutrophication can be described in terms of network attributes as any increase in system ascendency (due to nutrient enrichment) that causes a rise in total system throughput (TST) that more than compensates for a concomitant decline in the mutual information (Ulanowicz, 1986). This particular combination of changes in variables allows one to distinguish between instances of simple enrichment and cases of undesirable eutrophication.

Estimating a system's ascendancy implies the calculation of a set of information indices, followed by trophic and cycling analyses.

(1) Information indices

TST: The differences in system activity are gauged by the relative values of the TST. The TST is simply the sum of all transfer processes occurring in a given system. For all possible transfers T_{pq} that is:

$$TST = \sum_{p,q} T_{pq}$$

where p and q can represent either an arbitrary system component or the environment.

Ascendency: This is a key property of a network of flows that quantifies both the level of system activity and the degree of organization (constraint) with which material is being processed in autocatalytic systems such as ecosystems. The ascendency A, expressed in terms of trophic exchanges, T_{ij}, from taxon i to taxon j is calculated as:

$$A = \sum_i \sum_j T_{ij} \log \left[\frac{T_{ij} T_{..}}{T_{.j} T_{i.}} \right]$$

where a dot as a subscript indicates summation over that index.

Development Capacity: This index is a surrogate for the complexity of the food web (Monaco & Ulanowicz, 1997). In other words, it is the diversity of the system flows scaled by the TST. Quantitatively, it takes the form:

$$C = \sum_{i,j} T_{ij} \log \left[\frac{T_{ij}}{T_{..}} \right]$$

Average Mutual Information (AMI): Measures the average amount of constraint exerted upon an arbitrary quantum of currency as it is channelled from any one compartment to the next (Ulanowicz, 1997). It is the unscaled form of the ascendency and is written as:

$$AMI = \sum_{i,j} \frac{T_{ij}}{T_{..}} \log\left[\frac{T_{ij}T_{..}}{T_{i.}T_{.j}}\right]$$

Redundancy: This is the degree to which pathways parallel each other in a network. It can be calculated in an isolated system as the (non-negative) difference by which the system capacity exceeds the ascendency. In terms of flows it is:

$$R = -\sum_{i,j=0}^{n} \frac{T_{ij}}{T_{..}} \log\left[\frac{T_{ij}^2}{T_{i.}T_{.j}}\right]$$

where n is the number of components in the system (for more details see Ulanowicz & Norden, 1990; Ulanowicz & Wulff, 1991).

Specific Overhead of the system **(Ø/TST)**: It measures the total flexibility of the system on a per-unit-flow basis. The overhead of a system is the amount by which the capacity of a non-isolated system exceeds the ascendency. It consists mostly of redundancy, but in open systems it is also increased by multiplicities in the external inputs and outputs. In terms of flows, it resembles redundancy, except it also includes the transfers with the external world:

$$\frac{\Phi}{TST} - \sum_{i,j=0}^{n+2} \frac{T_{ij}}{T_{..}} = \log\left[\frac{T_{ij}^2}{T_{i.}T_{.j}}\right]$$

where the index $(n + 1)$ signifies an import and $(n + 2)$ an export or dissipation.

(2) Trophic analysis

Food webs that are qualitatively very different can be mapped into a standard straight-chain network topology. This standard form allows comparing corresponding trophic efficiencies between different estuaries (Baird et al., 1991). The trophic efficiency between any two levels is defined as the amount a given level passes on to the next one, divided by how much it received from the previous level (Ulanowicz & Wulff, 1991). Connectance indices (overall connectance and intercompartmental connectance) are estimates of the effective number of links both into and out of each compartment of a weighted network.

(3) Cycle analysis

The Finn Cycling Index (FCI) reveals the proportion of TST that is devoted to the recycling of carbon (Finn, 1976). Thus, $FCI = T_c/TST$, where T_c is the amount of system activity involved in cycling.

Patrício et al. (2004) applied ascendency to data on the Mondego estuarine intertidal communities showing that network analysis appeared to provide a systematic approach to apprehending what is happening at the whole-system level, which is obviously powerful from the theoretical point of view. Moreover, the study on the Mondego estuarine ecosystem provided an example of how the measures coming out of network analysis can lead to an improved understanding of the eutrophication process itself. Nevertheless, there is a major inconvenience regarding its use, that is the extremely considerable time and labour needed to collect all the data necessary to perform network analysis, which limits its application.

2.3.6.2   Eco-Exergy (Jørgensen & Mejer, 1979, 1981)

$$Ex = RT \times \sum_i C_i \times \beta_i$$

where R is the gas constant, T the absolute temperature, C_i the concentration of component i in the ecosystem (e.g. biomass of a given taxonomic group or functional group) and β_i is a factor able to roughly express the quantity of information embedded in the organisms' biomass, namely accounting for genome size (Table 13). Detritus was chosen as a reference level, that is $\beta_i = 1$ and Exergy in biomass of different types of organisms is expressed in detritus energy equivalents.

Table 13: Values for the number of genes and cell types and for the weighting factor (β) to estimate Exergy.

Primitive organisms	Plants	Animals	β-values
Detritus			1
Virus			1.01
Minimal cell			5.8
Bacteria			8.5
Archaea			13.8
Yeast			17.8
Protists	Algae		20
		Mesozoa, Placozoa	33
		Protozoa, Amoeba	39
		Phasmida (stick insects)	43
Fungi, moulds			61
		Nemertina	76
		Cnidaria (corals, sea anemones, jelly fish)	91
	Rodophyta		92
		Gastrotricha	97
Porifera			98

(Continued)

Table 13: (Continued)			
Primitive organisms	Plants	Animals	β-values
		Brachiopoda	109
		Platyhelminthes	120
		Nematoda	133
		Annelida	133
		Gnathostomulida	143
	Mustard weed		143
	Seedless vascular plants		158
		Rotifera	163
		Kinorhyncha	165
		Entoprocta	164
	Moss		174
		Insecta (beetles, flies, bees, wasps, bugs, ants)	167
		Coliodiea (sea squirt)	191
		Insects, Lepidoptera	221
		Crustaceans	232
		Chordata	246
	Rice		275
	Gymnosperms, including *Pinus*		314
		Mollusca, Bivalvia	310
		Mollusca, Gastropoda	310
		Insecta (Mosquito)	322
	Angiosperms		393
		Fish	499
		Amphibia	688
		Reptilia	833
		Aves	980
		Mammalia	2127
		Monkeys	2138
		Anthropoid apes	2145
		Homo sapiens	2173

Note: Values of weighting factors are based on the number of information genes. Estimations were carried out according to the method described by Jørgensen *et al.* (1995), based on analytical work (Fonseca *et al.*, 2000) and on literature sources (Lewin, 1994; Jørgensen *et al.*, 2005, 2007).

If the total biomass in the system remains constant, then Exergy variations will rely upon its structural complexity. Specific Exergy is defined as Exergy/ biomass. Both Exergy and Specific Exergy may be used as indicators in environmental management, it being advisable to apply them complementarily (Marques *et al.*, 1997).

This formulation of Exergy, referred in a first instance as Modern Exergy (Jørgensen *et al.*, 1995), does not correspond to the strict thermodynamic definition of the concept, but provides, nevertheless, an approximation of Exergy values. In this sense it was proposed to call it Exergy Index (Marques *et al.*, 1997, 1998), or Ecological Exergy (Fonseca *et al.*, 2000, 2002), a term finally adopted by Jørgensen as Eco-Exergy (see for instance Jørgensen *et al.*, 2005). This formulation allows an empirical estimation of the Exergy Index from normal sets of ecological data, for example organism's biomass, provided that the β_i value for the different types of organisms is known.

Marques *et al.* (1997) suggested the use of nuclear DNA (*C*-values) content to evaluate parameter β, assuming the DNA content is a measure of the information carried in its genome, acquired throughout the evolutionary process. On the other hand, Fonseca *et al.* (2000, 2002), in accordance with the studies by Lewin (1994), claim that organisms similar in complexity may have significantly different nuclear DNA content and that at higher evolutionary levels, genome size looses correspondence to the increase in structural complexity of organisms due to the presence of repetitive DNA sequences. Thus, non-repetitive DNA content, rather than the total genome should better evaluate organism complexity. Therefore, it could be assumed that to each adjacent triplet of nucleotides from non-repetitive DNA there is a corresponding RNA signal (from regulatory genes or structural genes). Hence, the non-repetitive DNA could be considered as an approximate estimation (although rough) of the overall 'coding capacity' of the genome and used in the evaluation of parameter β. For this reason, Fonseca *et al.* (2000) propose that, instead of using *C*-values to estimate weighing factor β for each species, the lowest (known) *C*-value in different groups of organisms is preferable. Either way, the estimation of correct β values constitutes one of the major difficulties involved in applying the Exergy concept in Ecology, and requires further research (Jørgensen *et al.*, 2005).

The minimum DNA contents (lowest *C*-values) of several groups used in the estimation of the β parameter are given in Table 13. In terms of Thermodynamics, Exergy applied in Ecology is a measure of the distance between a given state of an ecosystem and what the system would be like at a point of thermodynamic equilibrium (Jørgensen & Mejer, 1979). In other words, the Exergy of an ecosystem at thermodynamic equilibrium would be zero. This means that, during ecological succession, Exergy is used to build up biomass, which in turn stores Exergy, and therefore Exergy represents a measure of the biomass structure in addition to the information embedded in the biomass (Jørgensen *et al.*, 2002).

In a trophic network, biomass and Exergy will flow between ecosystem compartments, supporting different processes by which Exergy is both degraded (respiration) and stored (growth production) in different forms of biomass belonging to different trophic levels. More complex organisms have more built-in information and are further away from a thermodynamic equilibrium than simpler organisms. Therefore, more complex organisms also have more built-in Exergy (thermodynamic information) in their biomass than the simpler ones. On the other hand, ecological succession goes from more simple to more complex ecosystems, which seem to reach, at a given point, a sort of balance between maintaining a given structure, emerging for the optimal use of the available resources and modifying the structure, adapting it to a permanently changing environment.

Exergy has been considered a promising indicator of ecosystem integrity by several authors (Nielsen, 1990; Jørgensen, 1994; Fuliu, 1997), acquiring a considerable interest in the context of system ecology.

In fact, Exergy has been applied as an indicator of the state of ecosystems in a number of European lakes, mainly through the studies of Jørgensen (1994) and Nielsen (1992, 1994). Lakes have been investigated in connection with natural or human-induced changes in their ecosystems, such as eutrophication and biomanipulation. In addition, four other works investigated the relations between Exergy-based indices and biodiversity in a freshwater system, in an estuary, in a coastal lagoon and in an intertidal rocky shore, respectively (Jørgensen & Padisak, 1996; Marques *et al.*, 1997; Salas *et al.*, 2005; Patrício & Marques, 2006). Results showed that the Exergy-based indices appeared capable of providing useful information regarding the state of the systems.

2.4   Brief review of socio-economic indicators: the case of coastal environments

In general, human activities generate a series of damages and environmental stresses, which become evident in the alteration of natural processes that take place in different ecosystems. On the other hand, adding to environmental concerns, social, cultural and economical problems are overlapped, meaning that human activities are never isolated nor do they disturb the environment through cause–effect linear relations. Instead, they interact, meet and compete for the areas, summing up effects and producing a complex net of interrelations which makes it all the more difficult to analyse the situations.

The United Nations (UN) developed a list of environmental indicators in collaboration with the Inter-governmental Working Group on the Advancement of Environment Statistics, and following the fourth meeting, the Working Group (Stockholm, 6–10 February 1995) agreed on a list of environmental and related socio-economic indicators provided in Table 14.

On the other hand, regarding coastal environments, issue-specific global programmes such as the *Millennium Ecosystem Assessment* and the *World Commission on Protected Areas Marine Program*, which follow an integrated approach or perspective with a focus on ecosystems and on marine protected areas, respectively, have developed different socio-economic indicators. These

Table 14: Socio-economic activities and their impacts on the environment.

	Socio-economic activities and events	Impacts and effects<
Economic issues	Real GDP *per capita* growth rate	EDP/EVA per capita
	Production and consumption patterns	Capital accumulation (environmentally adjusted)
	Investment share in GDP	
Social/ demo- graphic issues	Population growth rate	% of urban population exposed to concentrations of SO_2, particulates, ozone, CO and Pb
	Population density	
	Urban/rural migration rate	Infant mortality rate
	Calorie supply per capita	Incidence of environmentally related diseases
Air/Climate	Emissions of CO_2, SO_2, and NO_x	Ambient concentrations of CO, SO_2, NO_x, O_3, and TSP in urban areas
	Consumption of ozone depleting substances	
		Air-quality index
Land/Soil	Land use change	Area affected by soil erosion
	Livestock per km^2 of arid and semi-arid lands	Land affected by desertification
	Use of fertilizers	Area affected by salinization and water logging
	Use of agricultural pesticides	
Fresh water resources	Industrial, agricultural and municipal discharges directly into freshwater bodies	Concentration of lead, cadmium, mercury and pesticides in fresh water bodies
	Annual withdrawals of ground and surface water	Concentration of faecal coliform in fresh water bodies
	Domestic consumption of water per capita	Acidification of fresh water bodies
	Industrial, agricultural water use per GDP	BOD and COD in fresh water bodies
		Water-quality index by fresh water bodies
		Deviation in stock from maximum sustainable yield of marine species
		Loading of N and P in coastal waters

(Continued)

Table 14: (Continued)		
	Socio-economic activities and events	Impacts and effects<
Marine water resources	Industrial, agricultural and municipal discharges directly into marine water bodies	Deviation in stock from maximum sustainable yield of marine species
	Discharges of oil into coastal waters	Loading of N and P in coastal waters
Biological resources	Annual round wood production	Deforestation rate
	Fuel wood consumption per capita	Threatened, extinct species
	Catches of marine species	
Mineral (incl. energy) resources	Annual energy consumption per capita	Depletion of mineral resources (% of proven reserves)
	Extraction of other mineral resources	Lifetime of proven reserves
	Municipal waste disposal	Area of land contaminated by toxic waste
	Generation of hazardous waste	
	Imports and exports of hazardous wastes	
Human settlements	Growth rate of urban population % of population in urban areas	Area and population in marginal settlements
		Shelter Index
		% of population with sanitary services
	Motor vehicles in use per 1000 habitants	
Natural disasters	Frequency of natural disasters	Cost and number of injuries and fatalities related to natural disasters

Source: http://unstats.un.org/unsd/ENVIRONMENT/indicators.htm

programmes look at both environmental and socio-economic aspects and their interactions.

UNESCO (2003) summarized these socio-economic indicators and those considered by national, state, local or site-specific coastal management programmes. In this summary, we can find indicators focusing on the following approaches:

(1) Coastal population (population density and population in coastal high hazard areas);
(2) Quality of life in the coastal zone (unemployment levels, perceived quality of coastal landscape, availability of affordable housing and population's age structure);
(3) Public-awareness information (public awareness of coastal issues and public awareness of sustainable development);
(4) Public access (legal availability, and access points);
(5) Service needs and provision (education, health, welfare, housing, water and sanitation, electricity, wastewater and stormwater, roads, railways, airports and harbours, telecommunication and postal services);
(6) Tourism and recreation (value of tourism and employment in the tourism sector, importance of tourism in the economy, tourist's arrivals, equitable distribution of tourism benefit);
(7) Fisheries (annual catch of major target species, percentage of household income deriving from fishing);
(8) Coastal community development (environment and land use, economic diversity and positive and negative economic growth, engagement between the government and the public, public investment and infrastructures);
(9) Development funding (level of finance from multilateral institutions and other institutional funding sources);
(10) Coastal dependent uses (description of the authority to enact laws and ordinances for the protection of public health, safety and welfare, and economic health measured by the different types and trends in economic development);
(11) Community participation (number of people involved in coastal management activities and extent of participation, level of awareness of coastal issues, participation of business and commercial enterprises in coastal management activities, participation in volunteer activities that protect and enhance coastal resources);
(12) Coastal hazards (population in coastal high hazards areas, emergency evacuation, shelter demand and capacity, level of awareness of coastal hazards, number of reported vessel incidents and land acquired for hazard mitigation);
(13) Waterfront revitalization (number of volunteers contributing time to activities associated with waterfront revitalisation, public and private investment in waterfront communities and number of community goals achieved).

3 Decision tree for selecting ecological indicators based on benthic invertebrate fauna data sets

3.1 Selecting ecological indicators: practical constrains

Theoretically, all ecological indicators accounting for species composition and abundance of biological communities might be useful in characterising the environmental situation of an ecosystem. However, since many indicators were developed to approach the characteristics of a specific ecosystem, they often lack generality. On the other hand, many have been criticized or rejected due to their dependence on specific environmental parameters, or because of their unpredictable behaviour depending on the type of environmental stress. Deciding on which set of ecological indicators to use in a particular case is therefore a sensitive process.

In the process of selecting an ecological indicator, the pollution type, the community type, data requirements and data availability must be accounted for. Moreover, the complementary use of different indicators or methods based on different ecological principles is highly recommended in determining the environmental quality status of an ecosystem (Dauer et al., 1993).

Since there are numerous kinds of pollution in the marine environment, one type of classification possible would be according to the pollutant nature. This can be done by considering the pollution caused by toxic substances, toxic thermal pollution, radioactive, organic and microbiological pollution, amongst others, all of which imply changes in the dynamics of the marine environment functioning without having a specific characterisation: suspended materials, fresh water input, etc. Among these types, organic and toxic pollution as well as physical disturbances are the most-frequent perturbations in coastal areas.

Pollution caused by toxic substances may affect the organisms' physiology negatively, possibly even leading to their death. Such substances are usually characterised as highly persistent and cause bioaccumulation in organisms via trophic networks. This provokes serious problems even at low concentrations, interfering in the enzymatic function and inducing the production of mixed-function oxidases (Kurelec et al., 1984; Spies et al., 1984; Hansen & Addison, 1990; Lafaurie et al., 1993) or metallothioneines (Harrison et al., 1988; Carpene, 1993; Roesijadi, 1994; Stewart, 1994; Ringwood et al., 1995). Among the toxic substances most frequently found in the marine environment are organochlorates, characterised by a high persistence to chemical and bacterial degradation, biphenyl-chlorates, dioxins, etc. Their presence in the sea is commonly associated with agricultural and industrial activities.

Heavy metals constitute another particular case of toxic pollution. They are released from mining drainage, industrial waste discharges, mud from sewage treatment plants and direct dumping deriving from metal application and transformation industries. Heavy metals can be adsorbed by mineral and organic particles that tend to drag them to the sea bottom where they remain, retained in the sediment, and are assimilated by living organisms that incorporate them in the trophic chains, causing bioaccumulation.

There is no doubt that organic pollution can be claimed as the most generalized and can be defined as the presence in the marine environment of organic substances that in principle are capable of being biodegraded and assimilated by coastal systems. The volume of organic matter dumped into our coasts exceeds that of any other pollutant, since any urban settlement, however small, is capable of producing large quantities of organic waste. Such pollution is usually associated with urban drainage network dumping, although agricultural activity, through the use of fertilizers, and other different kinds of industries, namely agroalimentary, paper and fish farming contribute in a very important way to the organic enrichment of the coastal environment.

On the other hand, the main physical stressors consist of mechanical disturbance (e.g. fishing), removal of substratum (e.g. aggregated or by dredging), changes in grain size or in temperature, suspended sediment, water-flow rates and sediment deposition (smothering).

Assessing the effect of any disturbance on the biological component of the ecosystem or community is often dependent on the type of data available. Moreover, applying different ecological indicators depends on a series of requisites.

Taking into account the indicators more frequently applied in marine and coastal studies, we provided a consensual decision tree (Figure 4) to be used in selecting the most-suitable indices or ecological indicators for each case, except for merely graphical methods due to their high subjectivity. We kept in mind the most-frequent types of disturbance occurring in coastal areas, namely organic enrichment, physical disturbance (mechanical disturbance and removal of substratum) or toxic pollution, the required level of taxonomic identification of organisms and the type of substrate.

We must highlight that a decision tree such as this is never concluded. For the sake of this work, we have included the indicators most used in the literature. However, new indices can be considered in the structure of the decision tree following the selection criteria proposed.

3.2 Decision tree

The decision tree for selecting the most-adequate indices or ecological indicators as a function of disturbance type and benthic invertebrate fauna data availability is presented as follow. Letters **A** to **M** correspond to ecological indicators' groups resulting from the decision-tree exercise (see Figure 4 and Table 15).

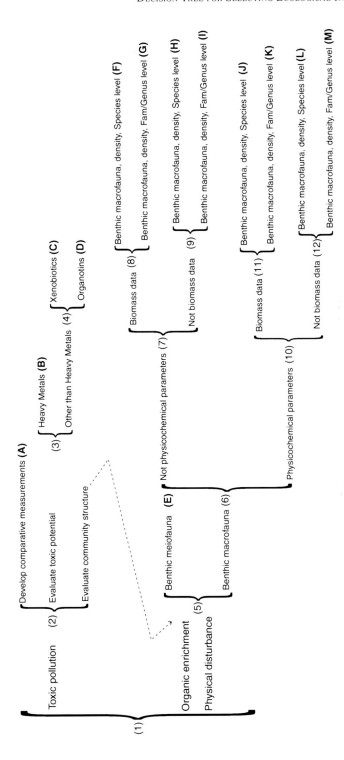

Figure 4: Decision tree for selecting ecological indicators based on benthic invertebrate fauna data sets (A–M. ecological Indicators' groups resulting from the decision-tree exercise, see Table 15).

1. The type of perturbation whose impact we intend to assess consists of
   toxic pollution.   2
   The type of perturbation whose impact we intend to assess consists of
   organic enrichment or physical disturbance.   5
2. The effect of pollution is to be evaluated at the level of the communities'
   structure.   5
   The effect of pollution is to be evaluated regarding the contaminant's
   toxic potential.   3
   The toxic reduction in organisms is determined with the sole intention of
   developing comparative measurements.   A
3. Contamination is caused by heavy metals.   B
   Contamination is caused by something other than heavy metals.   4
4. Contamination is caused by xenobiotics.   C
   Contamination is caused by organotins   D
5. Biological data available refers to a benthic meiofaunal community   E
   Biological data available refers to a benthic macrofaunal community   6
6. Our concern is the benthic macrofaunal community and the type of data
   available includes <u>no</u> information on physical–chemical parameters.   7
   Our concern is the benthic macrofaunal community and the type of data
   available includes information on physicochemical parameters.   10
7. The type of data available includes information on biomass of sampled
   organisms.   8
   The type of data available <u>does not</u> include information on biomass of
   sampled organisms.   9
8. Organisms are identified up to the species level   F
   Organisms are identified up to the genus or family level   G
9. The type of data available includes information on numerical density of
   sampled organisms and these are identified up to the species level   H
   The type of data available includes information on numerical density of
   sampled organisms and these are identified up to the genus or family level   I
10. The type of data available includes information on biomass of sampled
    organisms.   11
    The type of data available <u>does not</u> include information on biomass of
    sampled organisms.   12
11. Organisms are identified up to the species level   J
    Organisms are identified up to the genus or family level   K
12. The type of data available includes information on numerical density of
    sampled organisms and these are identified up to the species level   L
    The type of data available includes information on numerical density of
    sampled organisms and these are identified up to the genus or family
    level   M

Table 15: Ecological indicators' groups resulting from the decision-tree exercise.

	Ecological indicators
	Ecological indicators
A	Ecological reference index
B	ALA D inhibition (Pb) method, lysosomal neutral red retention method, lysosomal stability method, metallothionein induction method
C	Lysosomal neutral red retention method, lysosomal stability method
D	Imposex method, intersex method, shell-thickening method
E	Berger–Parker, indice of trophic diversity, Margalef, Maturity Index, Meiobenthic Pollution Index, Nematode/Copepod, Pielou, Shannon–Wiener, Simpson, Taxonomic Distinctness measures
F	*AMBI*, Annelida Pollution Index, APBI, Benthic Response Index, BEN-TIX, Berger–Parker, BOPA, Exergy, Feeding Structure Index, Fisher's α, Indicator Species Index, Infaunal Trophic Index, Hulbert, Macrofauna Monitoring Index, M-AMBI, Margalef, Molluscs Mortality Index, P-BAT, Pielou, Pollution Index, Polychaeta/Amphipoda Index, Shannon–Wiener, Simpson, Specific-Exergy, Taxonomic Distinctness measures, *W*-Statistic
G	APBI, Berger–Parker, Exergy, Fisher's α, Hulbert, M-AMBI, Margalef, Molluscs Mortality Index, P-BAT, Pielou, Polychaeta/Amphipoda Index, Shannon–Wiener, Simpson, Specific-Exergy, Taxonomic Distinctness measures, *W*-Statistic
H	*AMBI*, Annelida Pollution Index, APBI, Benthic Response Index, BEN-TIX, Berger–Parker, BOPA, Feeding Structure Index, Fisher's α, Hulbert, Indicator Species Index, Infaunal Trophic Index, Macrofauna Monitoring Index, M-AMBI, Margalef, P-BAT, Pielou, Pollution Index, Polychaeta/Amphipoda Index, Shannon–Wiener, Simpson, Taxonomic Distinctness measures
I	APBI, Benthic Response Index, Berger–Parker, Fisher's α, Hulbert, M-AMBI, Margalef, P-BAT, Pielou, Polychaeta/Amphipoda Index, Shannon–Wiener, Simpson, Taxonomic Distinctness measures
J	*AMBI*, Annelida Pollution Index, APBI, Benthic Condition Index, Benthic Quality Index, Benthic Response Index, BENTIX, Berger–Parker, B-IBI, BOPA, Coefficient of Pollution, Ecofunctional Quality Index, Exergy, Feeding Structure Index, Fisher's α, Hulbert, Indicator Species Index, Infaunal Trophic Index, Macrofauna Monitoring Index, M-AMBI, Margalef, Pollution Index, Molluscs Mortality Index, P-BAT, Pielou, Polychaeta/Amphipoda Index, Shannon–Wiener, Simpson, Specific-Exergy, Taxonomic Distinctness measures, VPBI, *W*-Statistic
K	APBI, Benthic Condition Index, Benthic Response Index, Berger–Parker Coefficient of Pollution, Ecofunctional Quality Index, Exergy, Fisher's α, Hulbert, M-AMBI, Margalef, Molluscs Mortality Index, P-BAT, Pielou, Polychaeta/Amphipoda Index, Shannon–Wiener, Simpson, Specific-Exergy, Taxonomic Distinctness measures, VPBI, *W*-Statistic

L	*AMBI*, Annelida Pollution Index, APBI, Benthic Condition Index, Benthic Quality Index, Benthic Response Index, BENTIX, Berger–Parker, BOPA, Coefficient of Pollution, Feeding Structure Index, Fisher's α, Hulbert, Indicator Species Index, Infaunal Trophic Index, Macrofauna Monitoring Index, M-AMBI, Margalef, P-BAT, Pollution Index, Polychaeta/Amphipoda Index, Pielou, Shannon–Wiener, Simpson, Taxonomic Distinctness, VPBI
M	APBI, Benthic Condition Index, Benthic Response Index, Berger–Parker, Coefficient of Pollution, Fisher's α, Hulbert, M-AMBI, Margalef, P-BAT, Pielou, Polychaeta/Amphipoda Index, Shannon–Wiener, Simpson, Taxonomic Distinctness, VPBI

4   Decision tree for selecting adequate indices or ecological indicators: examples of application

To illustrate the application of this decision tree for selecting the most-suitable indices or ecological indicators, as a function of the disturbance type and data availability, we used data from five study areas, two of them located in the Western coast of Portugal and the other three located in the South-Eastern coast of Spain.

Each study area represents a different disturbance scenario. The available data also proves them different in terms of their nature. In fact, while there is quantitative data for some areas, including comprehensive lists of species and their corresponding abundances and biomass, in other cases the only available data refers to the most-representative taxonomic groups in the community. Having such a set of heterogeneous raw data available presented an excellent opportunity to show how the exercise of applying the proposed key should be carried out.

4.1   Description of the study areas and the types of data available

4.1.1   The Mondego estuary

The Mondego estuary is located on the Western coast of Portugal and it consists of two arms, north and south, split by an island about 7 km from the sea and reconnecting near the mouth (Figure 5). Distinct hydrographic characteristics are found in the two arms. The northern arm is deeper than the southern arm, at present totally silted up in the upstream areas, causing the freshwater to flow essentially through the northern arm. Water circulation in the southern arm is dependent on tidal activity and on the small freshwater input from a tributary, the Pranto River, controlled by a sluice (Marques *et al.*, 2003). Harbour facilities and dredging activities, on the northern arm, cause physical disturbance at the bottoms, while freshwater discharges from agricultural areas in the river valley result in an excessive nutrient release into the southern arm (Marques *et al.*, 1993). Human pressure coupled with specific physical characteristics (water residence time, hydrodynamics and depth) and climate conditions (precipitation) have contributed to an environmental stress increase in the Mondego estuary (Dolbeth *et al.*, 2003). Nevertheless, and based on recent observation, the system appears to be recovering gradually from the effects of eutrophication that have aggressed it over the past two decades (Pardal *et al.*, 2004).

Figure 5: The Mondego estuary, Portugal.

Two different data sets were selected to estimate different ecological indicators:

(a) The first was provided by a study on the subtidal soft-bottom communities, which allowed characterising the whole system with regard to species composition and abundance, taking spatial distribution into account in relation to the physicochemical factors produced by both water and sediments. Benthic macrofauna was sampled twice during the spring, in 1990, 1992, 1998 and 2000, at 14 stations covering the entire terminal part of the system (Figure 6A);

(b) The second set proceeded from a study on intertidal benthic communities carried out from February 1993 to June 1994 in the southern arm of the estuary (Figure 6A). Samples of macrophytes, macroalgae and associated macrofauna, as well as samples of water and sediments, were taken. This was performed on a fortnightly basis at different sites, during low tide, along a spatial gradient of eutrophication symptoms, going from a non-eutrophic zone, where a macrophyte community (*Zostera noltii*) is present, up to a eutrophic one, in the inner areas of the estuary, from where macrophytes have disappeared while *Ulva* sp. (green macroalgae) blooms have been observed during the last decade. In this area, and following a pattern, *Ulva* sp. biomass normally increases from early winter (February/March) up to July, the period of time when an algal crash usually occurs. A second but much less-important algal biomass peak may sometimes be observed in September, followed by a decrease lasting until winter (Marques *et al.*, 1997).

In both studies, organisms were identified to the species level and their biomass was equally determined (g m^{-2}AFDW). Corresponding to each biological

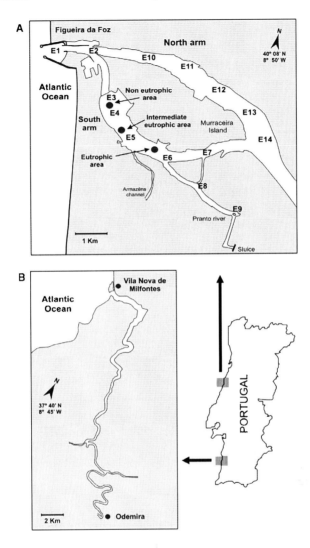

Figure 6: Portuguese case studies' locations. A: Mondego estuary, B: Mira estuary and sampling stations.

sample, the following environmental factors were determined: salinity, temperature, pH, dissolved oxygen, silica, chlorophyll *a*, ammonia, nitrates, nitrites and phosphates in water, as well as sediment organic-matter content.

4.1.2   The Mira estuary

The Mira estuary is 32-km long, extending from Vila Nova de Milfontes, close to the mouth, up to Odemira, at its upper limit (Figure 7). Both are small towns,

Figure 7:   The Mira estuary, Portugal.

constituting, nonetheless the most-important urban centres in that entire basin region. In general, the estuary is narrow and entrenched, with a width of approximately 150 m at its lower part and only 30 m in the upper reaches, with a mean depth of about 6 m (Costa *et al.*, 1994).

The Mira estuary and surrounding areas are included in the Natural Park of Sudoeste Alentejano and Costa Vicentina. The landscape is characterised by irrigated fields, well-preserved eucalyptus and cork-oak woods and their undergrowth (Raposo, 1996). The prevailing conditions allow, to a certain extent, considering the Mira estuary as representing a pristine system.

The selected data are included in the database developed by the TICOR Project (Bettencourt *et al.*, 2004), originating from a study operated by Andrade (1986) in which 99 sampling stations were utilized, covering the entire system (Figure 6B). Only benthic organisms' abundances are available.

4.1.3   The Mar Menor lagoon

The Mar Menor is a coastal lagoon with a total area of 135 km^2. The lagoon is connected to the Mediterranean at several points by channels through which water exchange takes place with the open sea (Figure 8).

The Mar Menor biological communities are adapted to more extreme temperatures and salinities than those found in the open sea. This coastal lagoon presents an environmental heterogeneity with different types of organic pollution too. Some areas are affected by: (a) direct urban dumping with the development of nitrophyle communities dominated by *Ulva* species; (b) dumping or zones

Figure 8 The Mar Menor coastal lagoon, Spain.

under the influence of harbours; (c) zones with high levels of sediment organic matter resulting from the primary production and the biological cycle of the macrophyte meadows (*Caulerpa prolifera*). This macrophyte was introduced into the lagoon as the result of dredging in one of the channels at the beginning of the 70s, growing rapidly throughout the lagoon, a phenomenon that has accelerated in the last years. Such *C. prolifera* growth has led to an increment of organic matter in the sediment and this increment, even if natural, had important consequences on the communities, with a general faunal impoverishment. In that sense, the named increment can be considered authentic pollution as it is understood by the GESAMP and (d) zones with low input of organic matter in the soft substrates (>1%) and on rocky bottoms.

To estimate different ecological indicators we used data collected by Pérez-Ruzafa (1989), since they have the advantage of constituting a complete characterization of the lagoon's benthic populations, containing all the information needed for a study such as the present one. Eleven sampling stations were located on rocky and soft substrates along the lagoon at sites representative of the different biocenosis and main polluted areas (Figure 9A). At some of the stations samples were taken seasonally (A: July, B: November, C: February, D: May) to evaluate the independence of different ecological indices with regard to seasonal variations.

The same happens with the Mondego estuary case study, where organisms were identified to the species level and their biomass was determined (g m^{-2}AFDW). The environmental factors taken into account were salinity, temperature, pH and dissolved oxygen, as well as sediment particle size, organic matter and heavy-metal contents.

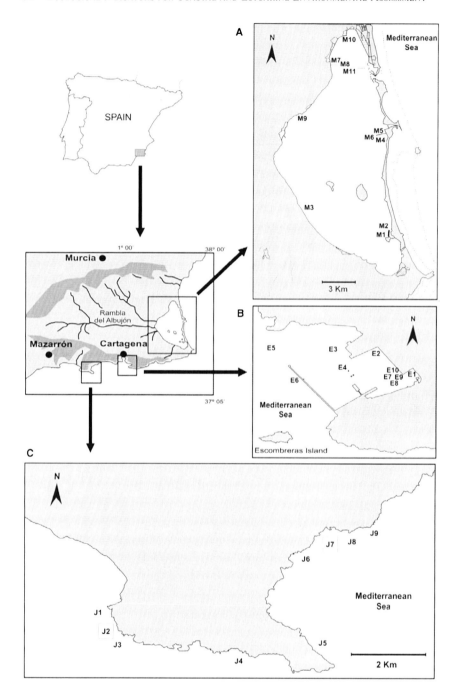

Figure 9: Spanish case studies' locations. A: Mar Menor coastal lagoon, B: Escombreras basin, C: Cape Tiñoso and sampling stations (squares around stations represent cages).

Figure 10: Escombreras basin. A: Aerial photograph; B: Industrial complex situated in the basin.

4.1.4   Escombreras basin

The Escombreras dock is located on the inlet of the same name, marked out by the Del Gate point in the northern extreme and by the Aguilones Point and the Escombreras Island in the South. This natural bay is closed by the Bastarreche dike-port and a series of greatly used docks and wharves can be found in the inner areas (Figure 10).

The Escombreras inlet, due to its own geographical characteristics and the proximity to the port city of Cartagena, has been developing itself as an important centre of maritime transport serving numerous factories situated in the Escombreras Valley. As a consequence, dumping and waste produced by these activities became a conditioning factor, which has been modelling the bay characteristics and the biological communities present there.

Most of the industrial and domestic waste matter produced in the area has been dumped in the marine environment; either directly into the bay or into places with close proximity and changes are visible. The marine communities as well as the physicochemical characteristics of both the waters and the marine sediments have been altered and modified in the last years due to such actions.

Data from the Escombreras dock were collected for an integrated study on pollution characterization carried out in the area in 1994 (Pérez-Ruzafa *et al.*, 1994).

Data were collected at 10 sampling stations (Figure 9B), in July 1994, describing the subtidal soft-bottom communities along the entire system with regard to species composition and abundance. A large number of pollutants proceeding from waste and industrial dumping could be identified in the area and these have led to an alteration of the physical and chemical characteristics of the waters and marine sediments.

4.1.5   Cape Tiñoso

Cape Tiñoso is located between La Azohía Point and Aguilones Point (Figure 11). The biocenosis in the area is characterised by a high specific diversity and a great

Figure 11: Cape Tiñoso. A: Aerial photograph; B: Floating cages assigned to red tuna fattening.

structural maturity, with a well-developed and preserved *Posidonia oceanica* meadow (C.A.R.M., 1998).

The high ecological value of the area is related with a low tourism pressure as compared to other zones of the Murcian littoral. Fishing is one of the only activities clearly developing from 1996 onwards, when the first floating cages assigned to red tuna fattening were installed, and have nowadays reached a considerable added value. The environmental effects of such an activity consist of (a) an increment in water turbidity, (b) an increment of water-dissolved nutrients and (c) direct sediments' organic enrichment due to faeces of the cultivated organisms and excess in food supply, with the corresponding ecological-related consequences.

Data from Tiñoso Cape were provided by a surveillance and monitoring environmental programme concerning fish-farming activities impact (Pérez- Ruzafa *et al.*, 1997). Nine sampling stations were established in the area assumed to be affected by the cages' influence, and two control stations in the eastern most extreme of Cape Tiñoso (Figure 9C).

One cage field is located between 37° 33′ 5″ N, 37° 33′ 25″ N and 1° 10′ 15″ E, near La Azohía point, about 20 to 40 m deep. A second field in the general area of the factory is located between 37° 33′ 50″ N, 37° 34′ 10″ N and 1° 6′ 5″ E y 1° 6′ 30″ E, near the Aguilar and the Bolete beaches, with a depth of 25–40 m. Sampling stations situated in the area under the influence of the cages were positioned in such a way that, with a single exception, all were in the course of the two dominant currents, the first with 0.38 knots of average speed, in the area of the first field of cages, and another with 0.33 knots of average speed, in the second field. Moreover, two sampling stations were located exactly under the cage fields (Figures 9C and 11).

At all stations, samples were taken in 1996 during the months of August and November, and in February, June and November of 1997. At each time, organic-matter content in the sediments, granulometry, as well as the concentration of nutrients in the water column were determined.

4.2   Indicator selection as a function of data availability

The first two things to bear in mind before applying the proposed key are: (a) the type of disturbance we want to measure and (b) the type of data we have. Table 16 provides a summary of the main disturbance factors and data availability regarding each one of these five case studies. If there happens to be no knowledge of disturbance type (like in the Mira estuary case), it is advisable to choose option 1 (organic enrichment) as it includes more non-pollution specific ecological indicators.

The next step is to select the most-appropriate indicators according to the available data nature, meaning data on organism abundance only or on organism abundance and biomass, when the species are identified up to species level, etc. Additionally, it must be taken into account that among indicators based on the

Table 16: Main disturbance factors and data availability regarding the five case studies.

Study area	Principal disturbance factor	Type of data
Mondego estuary	Organic enrichment	Abundance of benthic organisms
		Biomass of benthic organism (in the case of subtidal communities, data available only in 1998 and 2000)
		Physical–chemical parameters: (temperature, salinity, Chl a, nutrients, granulometry, % organic matter)
Mira estuary	Unknown	Abundance of benthic organisms
Mar Menor	Organic enrichment	Abundance of benthic organisms
		Biomass of benthic organisms
		Physical–chemical parameters: (temperature, salinity, granulometry, % organic matter, heavy-metal concentration in sediments)
Escombreras basin	Toxic pollution	Abundance of polychaeta species
		Biomass of polychaeta species
		Physical–chemical parameters: (temperature, salinity, nutrients, granulometry, % organic matter, heavy-metal concentration in sediments)
Cape Tiñoso	Organic enrichment	Abundance of polychaeta species
		Biomass of polychaeta species
		Physical–chemical parameters: (nutrients, Chl a, granulometry, % organic matter, heavy-metal concentration in sediments)

same principles, we should choose the ones which best include the characteristics that define a good ecological indicator.

For instance, among the indices based on indicator species, we have the Anellida Pollution Indice, the Bellan–Santini Pollution Index, the BENTIX, the Indicator Species Index and *AMBI*. Of this panoply, the most-appropriate index, since it is based on the classification of 3000 species and has been successfully tested in a higher number of geographical locations, would be *AMBI*. Adding to that, the fact that the authors provide free computer software for its application certainly makes it more than suitable. In fact, while BENTIX is overly specific for Mediterranean coastal waters, namely for areas near Greece, the Indicator Species Index is more specific to Norwegian and Swedish coastal waters. If we account for most of the multimetric indices, apart from being specific for given estuarine systems, they have mostly been developed for a concrete sampling area. Among those referred in the decision tree, Weisberg's B-IBI index is commonly used and, simultaneously, can easily be exported to other study areas. In Chapter 5, other multimetric indices will be applied and further discussed. Chapter 4 aims only to illustrate the application of the proposed decision tree. We do not intend to make an exhaustive exercise, calculating all indices referred into Chapter 2.

At a later stage, we will demonstrate how, with the help of the proposed decision tree, we can select the ecological indicators (indicated in bold) that can be applied in the five case studies considered, as a function of disturbance type, available data and, also, *a priori* knowledge regarding their characteristics.

Example 1:   The Mondego estuary

1. The type of perturbation whose impact we intend to assess
   consists of organic enrichment or physical disturbance. 5
5. Biological data available refers to benthic macrofaunal community 6
6. Our concern is the benthic macrofaunal community and the type of
   data available includes information on physicochemical parameters. 10
10. The type of data available includes information on biomass of
    organisms sampled. 11
11. Organisms are identified up to the species level ....**J**

J	*AMBI*, Annelida Pollution Index, APBI, Benthic Condition Index, Benthic Quality Index, Benthic Response Index, BENTIX, **Berger–Parker, B-IBI**, BOPA, Coefficient of Pollution, Ecofunctional Quality Index, **Exergy, Feeding Structure Index**, Fisher's α, Hulbert, Indicator Species Index, **Infaunal Trophic Index**, Macrofauna Monitoring Index, M-AMBI, **Margalef, Pielou**, Molluscs Mortality Index, P-BAT, Pollution Index, **Polychaeta/Amphipoda Index, Shannon–Wiener, Simpson, Specific Exergy, Taxonomic Distinctness measures**, VPBI, *W*-**Statistic**.

Example 2: The Mira estuary

1. The type of perturbation whose impact we intend to assess consists of organic enrichment or physical disturbance..5
5. Biological data available refers to benthic macrofaunal community..............6
6. Our concern is the benthic macrofaunal community and the type of data available includes <u>no</u> information on physicochemical parameters........7
7. The type of data available <u>does not</u> include information on biomass of organisms sampled..9
9. The type of data available includes information on numerical density of organisms sampled and organisms are identified up to the species level...... **H**

H	***AMBI***, Annelida Pollution Index, APBI, Benthic Response Index, BENTIX, **Berger–Parker**, BOPA, **Feeding Structure Index**, Fisher's α, Hulbert, Indicator Species Index, **Infaunal Trophic Index**, Macrofauna Monitoring Index, M-AMBI, **Margalef**, P-BAT, **Pielou**, Pollution Index, **Polychaeta/Amphipoda Index**, Shannon–Wiener, Simpson, **Taxonomic Distinctness measures**.

Example 3: The Mar Menor lagoon

1. The type of perturbation whose impact we intend to assess consists of organic enrichment or physical disturbance. 5
5. Biological data available refers to benthic macrofaunal community..............6
6. Our concern is the benthic macrofaunal community and the type of data available includes information on physicochemical parameters....10
10. The type of data available includes information on biomass of organisms sampled. 11
11. Organisms are identified up to the species level ....**J**

J	***AMBI***, Annelida Pollution Index, APBI, Benthic Condition Index, Benthic Quality Index, Benthic Response Index, BENTIX, **Berger–Parker**, **B-IBI**, BOPA, Coefficient of Pollution, Ecofunctional Quality Index, **Exergy**, **Feeding Structure Index**, Fisher's α, Hulbert, Indicator Species Index, **Infaunal Trophic Index**, Macrofauna Monitoring Index, M-AMBI, **Margalef**, Molluscs Mortality Index, P-BAT, **Pielou**, Pollution Index, **Polychaeta/Amphipoda Index**, Shannon–Wiener, Simpson, **Specific Exergy, Taxonomic Distinctness measures**, VPBI, ***W*-Statistic**.

Example 4: The Escombreras basin

1. The type of perturbation whose impact we intend to assess consists of toxic pollution. 2
2. The effect of pollution is to be evaluated at the level of communities' structure..5

J	***AMBI***, Annelida Pollution Index, APBI, Benthic Condition Index, Benthic Quality Index, Benthic Response Index, BENTIX, **Berger–Parker**, B-IBI, BOPA, Coefficient of Pollution, Ecofunctional Quality Index, Exergy, **Feeding Structure Index**, Fisher's α, Hulbert, Indicator Species Index, **Infaunal Trophic Index**, Macrofauna Monitoring Index, M-AMBI, **Margalef**, Molluscs Mortality Index, P-BAT, **Pielou**, Pollution Index, **Polychaeta/Amphipoda Index**, **Shannon–Wiener**, **Simpson**, Specific Exergy, **Taxonomic Distinctness measures**, VPBI, ***W*-Statistic**.

Example 5:   Cape Tiñoso

J	***AMBI***, Annelida Pollution Index, APBI, Benthic Condition Index, Benthic Quality Index, Benthic Response Index, Benthic Response Index, Benthic Response Index, BENTIX, **Berger–Parker**, B-IBI, BOPA, Coefficient of Pollution, Ecofunctional Quality Index, Exergy, **Feeding Structure Index**, Fisher's α, Hulbert, Indicator Species Index, **Infaunal Trophic Index**, Macrofauna Monitoring Index, M-AMBI, **Margalef**, Molluscs Mortality Index, P-BAT, **Pielou**, Pollution Index, **Polychaeta/Amphipoda Index**, **Shannon–Wiener**, **Simpson**, Specific Exergy, **Taxonomic Distinctness measures**, VPBI, ***W*-Statistic**.

4.3   Results of the application

4.3.1   The Mondego estuary

Firstly, the analysis was focused on the subtidal communities of both estuarine arms (first data set). Table 17 summarizes the indices' values obtained. Of all the utilized indicators, the only ones able to discriminate between different areas in

Table 17: Discrimination between different areas in the Mondego estuary based on the values estimated for several ecological indicators (Kruskal–Wallis test).

	AMBI	FSI	Margalef	TTD
	Average	Average	Average	Average
Northern arm-NDA	2.53	0.06	0.87	726.38
Northern arm-N	2.44	0.36	0.94	432.65
Southern arm-S	2.76	0.21	0.64	788.50
Southern arm-OM	3.08	0.19	0.32	610.12
GROUPS	1-NDA, N, S	1-NDA, S, OM	1-NDA, N	1-NDA,S, OM
	2-S-OM	2-N, S, OM	2-S, OM	2-NDA,N, OM
	3-OM			

Note: FSI, Feeding Structure Index; TTD, Total Taxonomic Distinctness; N, Non-dredged areas in the northern arm; DA, Dredged area in the northern arm; S, southern arm areas with organic-matter content < 5%; OM: Southern arm areas with organic-matter content > 5%.

the Mondego estuary are the Margalef index, the Total Taxonomic Distinctness (TTD) index, the Feeding Structure Index (FSI) and *AMBI* (Table 18). To be more precise, the Margalef Index and the TTD index, which are highly correlated ($r = +0.91$; $p < 0.001$) are only able to differentiate stations in the northern arm from those in the southern arm. The *AMBI* index, in its turn, can distinguish three groups: (a) stations in the southern arm with a higher percentage of sediment organic matter, which presented the highest values for the index (indicating greater disturbance), (b) stations affected by dredging activities and (c) less-disturbed stations of the northern arm. These last two groups are also differentiated by the FSI.

The discrimination of different areas by *AMBI* is fundamentally due to the dominance of the ecological groups III, IV and V in the southern arm stations presenting higher sediment organic-matter content (mainly stations E8 and E9). On the other hand, in the northern arm stations, species belonging to groups II and III are prevalent, although species from group IV started appearing since 1998. Regarding the northern arm, groups I, II and III are dominant in the stations less affected by environmental stress.

Despite the high correlation found between *AMBI* and B-IBI (Weisberg *et al.*, 1997) ($r = -0,61$; $p < 0.01$), the latter is not effective in discriminating between the different areas. *AMBI* values also appeared negatively correlated with Specific Exergy ($r = -0,67$; $p < 0.05$). This suggests that most of the information expressed by Specific Exergy was, in this case, very much related to the dominance of taxonomic groups usually absent in environmentally stressed situations.

Table 18: Indices' values in the Mondego estuary (subtidal) stations in 1990, 1992, 1998 and 2000.

	E1				E2				E3				E4			
	1990	1992	1998	2000	1990	1992	1998	2000	1990	1992	1998	2000	1990	1992	1998	2000
AMBI	–	–	1.2	1.53	2.62	1.58	1.87	3.22	2.11	1.90	3.00	2.45	2.86	2.50	2.76	2.39
FSI	–	–	0.75	1.67	0.33	0.40	0.25	0.25	0.33	0.25	0.40	0.40	0.13	0.20	0.20	0.17
ITI	–	–	81.4	95.96	79.70	59.05	45.85	51.52	50.08	70.82	33.34	54.55	62.91	74.08	60.25	56.67
Shannon–Wiener	–	–	2.85	0.87	1.56	1.74	2.45	3.45	2.56	2.94	0	2.55	3.11	2.42	1.43	2.92
Pielou	–	–	0.73	0.26	0.78	0.62	0.95	0.73	0.85	0.88		0.91	0.75	0.94	0.55	0.97
Margalef	–	–	2.32	1.30	0.91	1.26	1.35	4.01	1.47	2.01	0	1.74	3.15	1.47	0.94	1.99
Berger–Parker	–	–	0.43	0.87	0.62	0.66	0.30	0.25	0.34	0.23	1.00	0.36	0.41	0.33	0.73	0.20
Simpson	–	–	0.23	0.76	0.41	0.45	0.18	0.15	0.20	0.15	1.00	0.18	0.20	0.18	0.55	0.11
W-Statistic	–	–	0.27	-0.19		–	0.40	0.20	–	–	-1	0.45	–	–	-0.15	0.50
Δ	–	–	71.94	22.93	55.99	47.64	73.64	73.58	72.48	71.73	0	75.95	72.02	76.61	42.88	82.11
Δ	–	–	93.63	97.17	95.38	87.15	95.14	86.20	92.25	84.09	0	92.71	91.05	93.75	95.06	92.64
Δ+	–	–	92.38	85.56	97.22	83.33	83.33	90.51	90.48	87.04	0	91.27	90.69	91.11	94.44	92.26
TTD	–	–	1385.71	855.56	388.89	583.33	441.67	2353.33	633.33	870.37	0	638.89	1541.67	546.67	566.67	738.10
STTD	–	–	206.50	390.12	38.58	476.19	336.11	295.46	279.67	578.88	0	254.47	297.24	217.28	135.80	207.98
Exergy	–	–	214.08	3528.27		–	31.59	3424.53	–	–	5.76	15.04	–	–	6.53	31.09
Specific Exergy	–	–	99.75	276.30	–	–	218.84	217.43	–	–	450.00	65.44	–	–	159.35	348.10
B-IBI	–	–	3.67	3.67	–	–	2.67	3.00	–	–	2.33	3.00	–	–	1.67	2.67

	E5				E6				E7				E8			
	1990	1992	1998	2000	1990	1992	1998	2000	1990	1992	1998	2000	1990	1992	1998	2000
AMBI	3.08	3.18	3.04	3.04	3.04	3.07	2.84	2.99	2.00	3.07	3.00	3.05	3.25	3.15	3.08	3.07
FSI	0.29	0.11	0.13	0.11	0.13	0.29	0.17	0.13	0.22	0.33	0.17	0.11	0.71	0.14	0.17	0.17
ITI	66.67	64.65	58.57	67.88	65.49	61.68	63.77	68.99	66.67	66.67	57.31	69.79	63.02	67.25	60.68	66.42
Shannon–Wiener	1.22	2.74	2.03	2.51	0.72	1.88	1.91	1.46	1.61	1.44	1.66	2.39	1.93	2.35	1.47	1.68
Pielou	0.29	0.82	0.64	0.73	0.22	0.59	0.60	0.46	0.45	0.72	0.59	0.72	0.61	0.78	0.46	0.53
Margalef	2.11	1.91	1.07	1.35	1.18	1.32	1.25	1.03	1.57	0.71	0.81	1.43	1.25	1.18	0.98	1.15
Berger–Parker	0.81	0.33	0.61	0.37	0.90	0.54	0.43	0.69	0.66	0.65	0.61	0.42	0.61	0.34	0.70	0.46
Simpson	0.67	0.19	0.39	0.22	0.81	0.37	0.35	0.52	0.48	0.46	0.42	0.25	0.4	0.24	0.52	0.38
W-Statistic	–	–	−0.06	0.24	–	–	0.22	0.06	–	–	−0.04	0.11	–	–	−0.2	−0.09
Δ	28.19	76.88	49.52	67.82	16.03	61.11	51.87	46.73	42.56	51.6	56.22	69.71	47.72	72.55	40.85	47.63
Δ^*	84.71	94.71	81.73	87.43	86.75	96.39	81.17	96.38	81.94	95.91	97.62	93.18	84.42	95.27	85.4	78.44
Δ^+	87.15	91.48	88.43	88.89	92.59	86.57	89.88	86.57	93.64	94.44	92.86	90.74	89.88	94.05	94.64	94.05
TTD	1568.63	914.81	795.83	888.89	833.33	779.17	719.05	779.17	1030.00	377.78	650.00	907.41	719.05	752.38	757.14	752.38
STTD	343.13	205.21	305.86	308.64	207.48	475.61	284.51	444.74	217.08	154.32	173.85	266.12	304.35	182.82	139.95	182.82
Exergy	–	–	33.29	427.15	–	–	15.31	307.04	–	–	310.90	85.18	–	–	72.35	7.22
Specific Exergy	–	–	165.58	215.13	–	–	10.98	200.52	–	–	119.26	82.85	–	–	179.68	69.52
B-IBI	–	–	2.33	3.00	–	–	2.00	3.00	–	–	2.33	2.67	–	–	2.00	1.67

(Continued)

Table 18: (Continued)

	E9				E10				E11			
	1990	1992	1998	2000	1990	1992	1998	2000	1990	1992	1998	2000
AMBI	3.25	2.89	3.00	3.00	2.49	2.81	3.15	3.81	2.39	3.00	2.14	2.35
FSI	0.17	0	0	0.33	0.50	0	0	0.43	0	0.17	0	0.33
ITI	60.07	64.98	62.5	68.63	66.42	63.75	33.34	61.43	33.34	56.25	37.69	52.39
Shannon–Wiener	2.31	0.35	0.83	1.38	1.50	0.96	1.36	2.40	1.85	2.45	2.96	1.84
Pielou	0.73	0.22	0.36	0.59	0.95	0.42	0.68	0.72	0.92	0.82	0.89	0.92
Margalef	1.37	0.38	0.72	0.8	0.77	0.72	0.89	1.53	0.95	1.38	1.99	0.9
Berger–Parker	0.46	0.95	0.86	0.69	0.50	0.83	0.70	0.40	0.43	0.38	0.30	0.43
Simpson	0.27	0.9	0.74	0.51	0.32	0.69	0.50	0.25	0.27	0.23	0.15	0.28
W-Statistic	–	–	-0.18	0.24	–	–	0.21	0.23	–	–	0.59	0.39
Δ	62.18	10.09	24.59	42.66	63.13	25.28	47.65	44.5	59.72	69.56	74.27	67.70
Δ	85.33	99.39	96.1	91.53	93.25	81.66	95.83	59.22	100.00	90.08	87.54	94.12
Δ+	89.81	83.33	95.00	91.67	88.89	86.67	83.33	85.19	100.00	90.48	86.3	97.22
TTD	808.33	250.00	475.00	366.67	266.67	433.33	333.33	851.85	300.00	723.81	862.96	388.89
STTD	312.93	555.56	113.89	162.04	246.91	266.67	277.78	410.15	0	246.6	373.94	38.58
Exergy	–	–	3.13	1.67	–	–	21.39	4.52	–	–	3416.39	1.95
Specific Exergy	–	–	146.37	1.82	–	–	122.61	50.9	–	–	230.27	220.86
B-IBI	–	–	1.67	2.33	–	–	2.00	1.33	–	–	4.00	2.67

	E12				E13				E14			
	1990	1992	1998	2000	1990	1992	1998	2000	1990	1992	1998	2000
AMBI	0.29	–	2.25	2.5	2.61	1.87	2.24	1.20	2.87	2.96	1.65	3.03
FSI	0	–	0	0	0.1	0	0		0	0.50	0	–
ITI	57.25	–	47.51	33.34	47.15	33.34	46.38	53.34	59.56	92.89	36.37	58.12
Shannon–Wiener	1.56	–	2.14	0.65	2.95	1.75	2.61	1.37	1.20	0.55	0.87	2.04
Pielou	0.60	–	0.76	0.65	0.82	0.88	0.87	0.86	0.52	0.35	0.55	0.73
Margalef	0.99	–	1.26	0.27	1.95	0.73	1.55	0.67	0.79	0.36	0.60	1.23
Berger–Parker	0.65	–	0.37	0.83	0.25	0.44	0.39	0.60	0.77	0.89	0.82	0.54
Simpson	0.47	–	0.26	0.72	0.16	0.32	0.21	0.41	0.60	0.81	0.67	0.34
W-Statistic	–	–	0.3	–0.5	–	–	–0.05	0.20	–	–	0.18	0.19
Δ	39.49	–	69.02	28.48	78.67	65.28	73.57	58.92	35.36	19.30	32.58	57.11
Δ	74.08	–	95.35	100.00	93.74	96.30	92.55	100.00	88.75	99.57	100.00	86.20
Δ^+	76.67	–	84.44	100.00	88.79	94.44	89.29	100.00	90.00	88.89	100.00	89.68
TTD	460.00	–	506.67	200.00	976.67	377.78	714.29	300.00	450.00	266.67	300.00	627.78
STTD	288.89	–	424.69	0	303.58	154.32	321.71	0	233.33	246.91	0	224.24
Exergy	–	–	59.59	2.48	–	–	6.33	2.30	–	–	3.55	16.16
Specific Exergy	–	–	59.13	321.82	–	–	202.32	145.61	–	–	222.38	175.20
B-IBI	–	–	2.33	2.33	–	–	2.33	3.00	–	–	2.33	1.67

Note: FSI, Feeding Structure Index; ITI, Infaunal Trophic Index; Δ, Taxonomic Diversity; Δ^*, Taxonomic Distinctness; Δ^+, Average Taxonomic Distinctness; TTD, Total Taxonomic Distinctness; STTD, Variation in Taxonomic Distinctness.

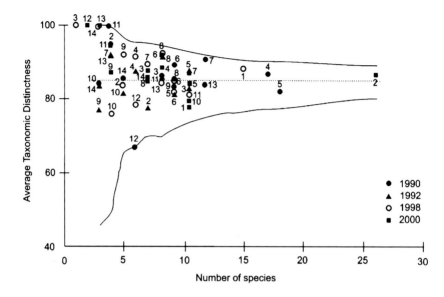

Figure 12: Confidence funnel (mean and 95% confidence interval) of the Average Taxonomic Distinctness in the Mondego estuary's subtidal stations (numbers above the symbols correspond to station number) in 1990, 1992, 1998 and 2000.

It becomes clear that most of the sampling stations do not show differences when accounting for the Average Taxonomic Distinctness values (Figure 12). In fact, even in the stations where only a small amount of species were observed (e.g. E12 and E13 in 2000; E14 in 1998), the Average Taxonomic Distinctness measures present higher values, suggesting therefore high path length between species throughout the tree.

None of the indicators was able to register significant differences between different years. Nevertheless, they all indicate an improvement in the environmental status in 2000, which coincided with the implementation of mitigation processes in practice since 1998.

Regarding the link between the physicochemical environmental factors and the variation of ecological indicator values, the Berger–Parker Index, the Exergy Index and the Average Taxonomic Distinctness index were the only indices sensitive to the parameters normally associated with eutrophication (Table 19).

Let us now considerer the macrobenthic intertidal communities along the eutrophication symptoms gradient in the Mondego estuary's southern arm. In this case, only the TTD index was able to significantly discriminate between the three areas considered, exhibiting higher values at the Z. noltii beds and the lowest at the most-eutrophic area (Table 20).

Furthermore, the Margalef Index and the Exergy Index behaved as expected, showing higher values in the Z. noltii area and lower values in the inner areas of

Table 19: Pearson's correlations between the indicators' values and physico-chemical parameters in the Mondego estuary (subtidal communities). $(*) = p < 0.05$.

	NO_2	NO_3	PO_4^{2-}	NH_4^+	Chl a
Berger–Parker	0.27	0.60	0.45	0.95*	−0.20
Average Taxonomic Distinctness	0.77*	0.43	0.63*	−0.20	−0.77*
Exergy	−0.68*	−0.67*	−0.80*	−0.36	0.48

Table 20: Discrimination between different intertidal areas along a gradient of eutrophication symptoms in the southern arm of the Mondego estuary, based on the values estimated for ecological indicators (Kruskal–Wallis test).

		TTD		Exergy		Margalef
		Average		Average		Average
Non-Eutrophic Area		2348.68		35,048.9		2.29
Intermediate Eutrophic Area		1919.39		10,143.89		2.08
Eutrophic Area		1542.78		14,893.58		1.60
Groups		1-NEA		1-NEA		1-NEA
		2-IA		2-IA,EA		2-IA, EA
		3-EA				

Note: TTD, Total Taxonomic Distinctness; NEA, Non-Eutrophic Area; IA, Intermediate Eutrophic Area; EA, Eutrophic Area.

the southern arm, although they did not allow discriminating the intermediate eutrophic area from the most-eutrophic one. On the contrary, the values estimated for several of the other indicators appear to indicate a better environmental status in the eutrophic area, which is inconsistent with our current knowledge of the system (Figure 13).

In this case, *AMBI* was unable to discriminate between the three areas, presenting in all these cases values close to 3, indicating slightly polluted scenarios, *sensu* Borja *et al.* (2000), where species of the ecological group III are dominant.

Exceptionally, *AMBI* values between 4 and 5 were estimated from July to October (Figure 13), in the intermediate eutrophic area, which indicates an average pollution situation. The Polychaete/Amphipod Ratio was able to illustrate the existing eutrophication gradient, exhibiting lower values in the *Z. noltii* beds and higher values at the intermediate and most-eutrophic areas, but was not sensitive enough to distinguish between the latter ($p \leq 0.05$).

Regarding correlations between the values of ecological indicators and physicochemical environmental factors, it is clear that the ITI index is highly correlated

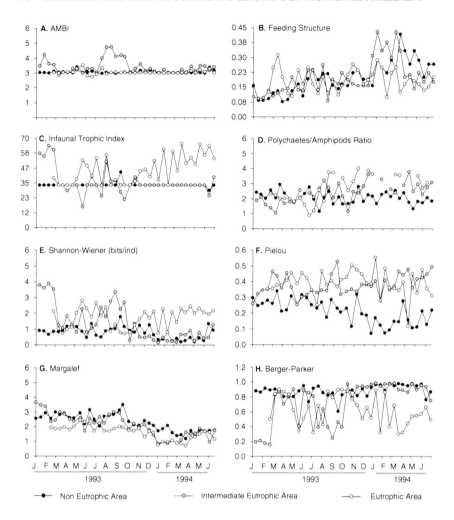

Figure 13: Temporal and spatial variation of the different ecological indicators applied to data on the intertidal communities of the southern arm of the Mondego estuary.

with the sediment organic-matter content ($r = -65$; $p < 0.001$), as was verified by Word (1990). Consequently, it presents the lowest values at the *Z. noltii* area, where the sediment organic-matter content (M.O.%) is higher (6.25% in average), indicating organic enrichment, and the highest values in the eutrophic area, where the percentage of organic matter in the sediments is lower (3.25% in average). Therefore, it must be concluded that this indicator is not sensitive to the process of eutrophication, responsible for the occurrence of *Ulva* sp. blooms and the decrease of the *Z. noltii* meadows.

Nevertheless, and despite being capable of differentiating the *Z. noltii* beds from the other areas in the southern arm of the estuary, the TTD index, as well as

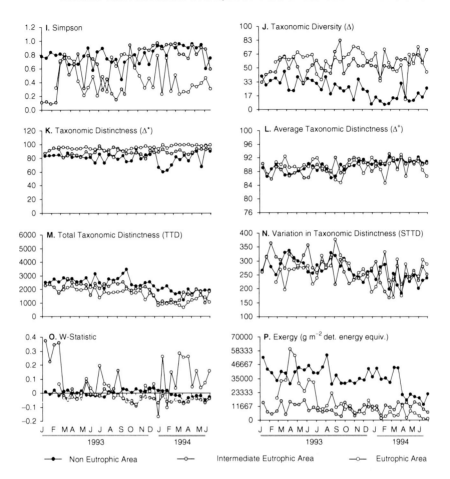

Figure 13 (Continued)

Table 21: Pearson's correlations between ecological indicators' values and the organic-matter content in the sediments for the intertidal communities of the Mondego estuary.

	TTD	Margalef	Pielou	Exergy
Organic-Matter Content (%)	−0.76**	−0.64*	−0.59*	−0.62*

Note: TTD, Total Taxonomic Distinctness. (*), $p < 0.05$; (**), $p < 0.001$.

the Margalef, Pielou and the Exergy-based indices are correlated positively and significantly to the sediment organic-matter content (Table 21). Moreover, the Exergy and the Margalef indices are negatively correlated with the ammonium and nitrite concentrations in the water column, respectively (Exergy against the ammonium concentration: $r = -0.30, p < 0.05$; Margalef Index against the Nitrites concentration: $r = -0.25, p < 0.05$). They demonstrate sensitivity to the fact that, in this case, the benthic communities' structure is strongly influenced by the water column nutrient concentration, but not so much by the sediment organic-matter content.

4.3.2   The Mira estuary

Regarding the Mira estuary, the absence of environmental data did not permit establishing any relation between the ecological indicators values and the physicochemical parameters. In addition, we could not apply ANOVAto assess the different indicators' performance in the different areas of the estuary, due to the fact that we could not set up any zonation criteria. The confirmation of this is presented through the application of the nMDS analysis (Figure 14).

In fact, Andrade (1986) had already mentioned the impossibility of distinguishing contrasting areas, considering the Mira estuary as a continuum. However, Pearson's correlations between the different ecological indicators results show interesting outcomes (Table 22).

Firstly, and as expected, significant correlations between the values of the Shannon–Wiener, Margalef, Simpson, Berger–Parker and Pielou indices were observed. Furthermore, Taxonomic Diversity was also highly correlated with

Figure 14:   Mira estuary. Two-dimensional MDS plot of taxa abundance (Clarke, 1993; Clarke & Warwick, 1994) (numbers above the symbols correspond to the station's number).

Table 22 Pearson correlations between the values of the different ecological indicators estimated based on data proceeding from sampling stations in the Mira estuary.

	AMBI	Berger	Δ+	Δ*	Δ	FSI	ITI	STTD	Margalef	Pielou	P/A	TTD	Shannon
Berger	0.20												
Δ+	-0.08	0.03											
Δ*	-0.24	-0.22	0.60*										
Δ	-0.24	-0.96*	0.12	0.43									
FSI	-0.51	0.06	-0.06	-0.06	-0.05								
ITI	0.03	0.24*	0.16	0.28*	-0.15	-0.01							
STTD	-0.22	-0.13	-0.48	-0.33*	0.07	0.21	-0.13						
Margalef	-0.10	-0.24	-0.18	-0.27*	0.19	0.31	-0.29	0.34*					
Pielou	-0.14	-0.88*	0.02	0.33*	0.89*	-0.13	-0.14	-0.09	-0.07				
P/A	0.20	0.28	-0.09	-0.54*	-0.36*	0.07	-0.38	0.16	0.47*	-0.51*			
TTD	0.02	0.02	-0.13	-0.36*	-0.07	0.27	-0.23	0.30*	0.93*	-0.35*	0.66*		
Shannon	-0.24	-0.88*	-0.04	0.15	0.87*	0.12	-0.24*	0.27*	0.58*	0.66*	-0.10	0.34*	
Simpson	0.21	0.98*	0.02	-0.25	-0.98*	0.02	0.23	-0.15	-0.27	-0.89*	0.30	0.00	-0.91*

Note: Δ, Taxonomic Diversity; Δ*, Taxonomic Distinctness; Δ+, Average Taxonomic Distinctness; AMBI, Taxonomic Distinctness (presence/absence of species); FSI, Feeding Structure Index; ITI, Infaunal Trophic Index; TTD, Total Taxonomic Distinctness; STTD. Variation in Taxonomic Distinctness; P/A, Polychaete/Amphipod Ratio. (*), $p \le 0.05$; (**), $p \le 0.01$.

Shannon–Wiener, Pielou and Simpson indices. This last correlation is expected to occur when the taxonomic tree falls into a single hierarchy level (all species of the same genus), and in that case, taxonomic diversity takes the form of the Simpson diversity (Warwick & Clarke, 1994).

On the other hand, TTD appears strongly correlated with the Margalef Index, as observed in the Mondego estuary. In fact, these two indices are based on species richness, and thus, TTD is supposed to be capable of differentiating when an assemblage consisting of closely related species is less rich than the one composed of distantly related species.

However, such a high correlation leads us to think that this is a case of an analogue performance of both indices, and therefore, in this case study, TTD cannot be considered better than the Margalef index when measuring diversity. It is also intriguing to discover that *AMBI* is correlated with Taxonomic Diversity, but not with any other diversity index. In fact, *AMBI* shows values below 2.5 in all cases, which indicates a good ecological status in all the stations (Table 23).

In the same way, as shown in Figure 15, the comparison of samples with the master list (TAXDTEST) showed that the Average Taxonomic Distinctness was, in most of the stations, within the 95% confidence intervals of the probability funnel for all samples, also indicating a good ecological status *sensu* Somerfield *et al.* (2003). These results match the assessment reached in a number of previous

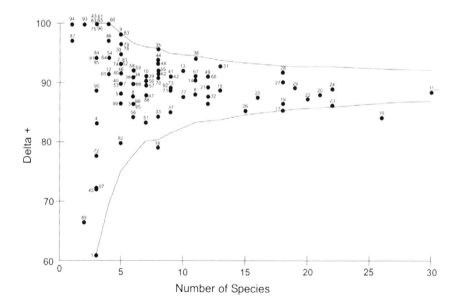

Figure 15: Departure from the theoretically expected Average Taxonomic Distinctness and 95% confidence funnel of all individual samples using presence/absence data of the Mira estuary (numbers above the symbols correspond to the station's number).

Table 23: Indices' values estimated based on data from each sampling station at the Mira estuary.

	AMBI	FSI	ITI	P/A	Shannon	Pielou	Margalef	Berger	Simpson	Δ	Δ*	Δ+	TTD	STTD
S1	1.29	0	33.34	—	1.06	0.67	0.67	0.75	0.57	22.48	52.56	61.11	183.33	432.10
S2	1.69	0.33	57.15	—	2.16	0.93	1.34	0.38	0.21	71.27	90.28	93.33	466.67	177.78
S3	1.05	0	33.34	—	2.37	0.92	1.55	0.30	0.19	70.83	87.18	86.67	520.00	340.74
S4	1.50	0	33.34	0.48	1.58	1.00	0.99	0.33	0.23	64.10	83.33	83.33	250.00	555.56
S5	1.75	0	33.34	—	2.25	0.97	1.61	0.33	0.15	72.73	85.71	88.33	441.67	336.11
S6	1.35	0.67	46.18	0.50	2.96	0.86	2.16	0.21	0.14	76.12	88.92	88.18	970.00	390.63
S7	0.12	2.00	57.15	—	1.87	0.81	1.13	0.43	0.31	51.65	74.49	90.00	450.00	233.33
S8	0.43	1.50	42.34	1.04	1.59	0.61	1.06	0.67	0.47	45.84	86.96	87.78	526.67	424.69
S9	0.68	0.25	60.61	—	2.12	0.91	1.21	0.36	0.23	74.90	97.04	98.33	491.67	25.00
S10	0	1.50	74.36	0	2.41	0.86	1.63	0.38	0.21	71.85	91.50	91.27	638.89	228.02
S11	1.31	0.50	58.73	0.94	4.04	0.82	5.50	0.24	0.1	77.82	86.39	88.58	2657.47	323.02
S12	0	1.00	35.56	—	1.20	0.60	0.62	0.62	0.5	49.08	97.24	91.67	366.67	347.22
S13	0.17	1.50	62.42	—	2.45	0.74	1.86	0.46	0.27	56.36	76.71	92.22	922.22	254.32
S14	0.19	0.25	33.34	0.88	1.49	0.43	1.82	0.76	0.59	39.71	96.76	90.61	996.67	330.95
S15	2.71	0.20	45.37	1.60	2.69	0.73	2.07	0.33	0.22	72.40	92.62	88.89	1155.56	318.14
S16	2.68	0.63	67.44	2.28	2.34	0.50	3.78	0.61	0.4	49.30	81.78	84.26	2190.67	460.69
S17	2.35	0.50	58.91	2.42	1.82	0.44	2.47	0.71	0.52	38.62	80.12	85.51	1539.22	311.16
S18	2.67	1.00	78.85	—	2.26	0.75	1.26	0.34	0.24	64.54	85.20	79.17	633.33	508.43
S19	2.33	0.42	64.35	1.25	3.10	0.74	3.16	0.31	0.17	66.37	80.08	86.71	1560.78	473.35
S20	2.68	0.46	96.67	—	1.48	0.34	2.80	0.77	0.6	32.83	82.73	88.17	1851.67	382.65
S21	1.56	0.20	33.34	—	2.67	0.74	2.36	0.36	0.22	68.01	87.00	87.88	1054.55	374.96
S22	2.22	0.11	55.78	0.83	2.45	0.57	2.65	0.53	0.32	59.23	86.59	87.37	1747.37	445.71

(Continued)

Table 23: (Continued)													
S23	2.54	0.83	63.62	2.28	2.24	0.50	3.29	0.64	0.43	49.07	85.60	86.36	1900.00	426.12
S24	2.85	0.43	66.67	3.00	0.96	0.22	2.66	0.88	0.77	19.47	84.84	89.11	1960.32	262.50
S25	2.70	0.60	66.67	—	1.18	0.29	2.18	0.83	0.69	23.02	73.56	87.64	1402.22	418.96
S26	2.72	0.25	66.67	—	0.60	0.15	1.62	0.90	0.81	13.00	69.30	85.40	1280.95	358.18
S27	2.85	0.20	66.67	2.58	0.52	0.12	1.94	0.93	0.87	10.08	76.37	90.31	1625.49	292.72
S28	2.72	0.13	66.67	2.34	0.84	0.20	2.07	0.88	0.77	16.66	72.81	91.94	1654.90	254.56
S29	2.78	0.20	66.67	2.57	0.69	0.16	2.18	0.91	0.83	13.27	76.61	89.28	1696.30	300.91
S30	1.20	0.25	62.82	—	2.13	0.92	0.90	0.32	0.23	75.95	98.95	95.00	475.00	225.00
S31	2.65	0.18	66.67	2.44	0.79	0.21	1.46	0.87	0.76	17.44	72.56	92.95	1208.33	203.13
S32	2.80	0	66.67	2.92	0.68	0.19	1.39	0.90	0.82	13.38	73.08	86.62	1039.39	355.38
S33	2.57	0	60.56	2.65	0.98	0.33	0.96	0.81	0.67	22.70	67.96	84.52	676.19	415.25
S34	1.82	0	65.51	1.96	1.84	0.71	0.84	0.51	0.34	56.33	85.38	91.11	546.67	217.28
S35	0.92	0.33	86.36	1.49	1.94	0.65	1.13	0.46	0.32	66.69	98.75	95.83	766.67	111.61
S36	1.50	0.38	58.49	0.60	2.85	0.82	2.05	0.38	0.19	74.93	92.86	94.24	1036.67	143.62
S37	1.89	0.29	92.93	0.11	2.42	0.76	1.56	0.36	0.23	67.52	88.13	85.19	766.67	582.99
S38	1.54	0	57.80	1.53	2.17	0.84	1.01	0.36	0.24	62.31	82.30	91.11	546.67	217.28
S39	1.55	0.75	37.85	1.27	0.60	0.21	0.99	0.92	0.84	14.70	92.25	90.48	633.33	306.12
S40	2.29	0	66.97	0.34	1.71	0.74	0.85	0.44	0.35	60.92	93.82	90.00	450.00	233.33
S41	1.38	0	61.17	0.66	2.51	0.79	1.58	0.37	0.23	71.34	92.80	91.20	820.83	254.42
S42	2.66	0.13	69.98	1.36	2.65	0.83	1.58	0.37	0.21	71.93	90.49	91.20	820.83	300.71
S43	2.63	0	89.82	0.08	0.88	0.55	0.37	0.80	0.66	33.58	100.0	100.0	300.00	0
											0	0		
S44	0.89	0.60	76.49	—	2.65	0.88	1.50	0.31	0.18	76.50	93.27	94.05	752.38	143.14

S45	3.00	0	37.19	—	1.41	0.89	0.61	0.50	0.38	50.87	82.46	72.22	216.67	1543.21
S46	2.23	0	73.94	0.21	1.43	0.62	0.80	0.58	0.44	53.29	95.89	91.67	458.33	291.67
S47	2.13	0.17	73.50	0.23	2.06	0.73	1.34	0.42	0.3	64.50	92.49	88.10	616.67	506.42
S48	2.37	0.14	54.50	0.78	2.65	0.88	1.69	0.37	0.19	71.96	88.77	93.45	747.62	224.99
S49	1.83	0.20	77.28	0.52	3.03	0.85	2.05	0.25	0.15	80.91	95.39	91.16	1093.94	296.46
S50	1.76	0	63.69	0.37	2.37	0.85	1.49	0.35	0.23	72.32	93.33	90.48	633.33	253.21
S51	2.61	0	61.42	0.37	2.36	0.84	1.24	0.37	0.23	67.36	87.73	83.33	583.33	529.10
S52	2.55	0	90.79	0.08	1.30	0.56	0.67	0.74	0.56	42.69	97.35	93.33	466.67	177.78
S53	1.70	0	78.01	0.30	2.16	0.93	0.80	0.39	0.25	72.08	95.59	90.00	450.00	233.33
S54	2.18	0	88.33	0.12	1.44	0.72	0.61	0.62	0.44	55.35	99.52	94.44	377.78	154.32
S55	2.29	0.14	91.35	0.05	1.65	0.55	1.25	0.66	0.47	50.51	95.57	92.26	738.10	227.82
S56	1.15	0.20	84.33	0.36	1.97	0.76	0.94	0.47	0.31	66.38	96.42	84.44	506.67	720.99
S57	2.65	0	60.26	0.34	2.26	0.81	1.19	0.38	0.25	66.70	88.73	89.68	627.78	330.06
S58	2.35	0	82.93	0.14	1.63	0.63	1.04	0.65	0.45	53.07	97.12	92.22	553.33	180.25
S59	2.14	0	45.90	0.73	2.13	0.82	1.18	0.48	0.28	63.49	88.46	92.22	553.33	217.28
S60	2.00	0	50.01	0.60	1.73	0.86	0.81	0.50	0.33	66.95	100.0	100.0	400.00	0
											0	0		
S61	2.51	0.50	96.52	0.03	1.11	0.70	0.41	0.67	0.53	46.98	100.0	100.0	300.00	0
											0	0		
S62	1.71	0.33	86.97	0.36	1.66	0.55	1.17	0.71	0.51	47.77	97.68	91.67	733.33	228.17
S63	3.00	0	57.54	—	1.08	0.68	0.40	0.62	0.51	49.11	100.0	100.0	300.00	0
											0	0		
S64	1.85	0.33	98.73	0.02	0.92	0.46	0.48	0.75	0.62	38.21	99.75	94.44	377.78	154.32

(Continued)

Table 23: (Continued)

S65	2.60	0	90.67	0.07	0.91	0.57	0.51	0.80	0.66	34.47	100.0	100.0	300.00	0
											0	0		
S66	2.98	0	55.65	1.04	1.48	0.57	0.89	0.60	0.45	53.64	98.33	90.00	540.00	325.93
S67	1.50	0.1	79.29	0.73	2.05	0.59	1.96	0.57	0.38	60.19	96.79	91.21	1003.33	321.76
S68	1.03	0.38	81.60	0.09	2.46	0.69	2.09	0.40	0.26	69.66	94.46	89.39	1072.73	333.64
S69	2.67	0.33	93.10	0.07	1.45	0.72	0.88	0.67	0.46	52.36	97.62	91.67	366.67	347.22
S70	1.57	0.33	98.62	0.35	0.53	0.18	1.02	0.93	0.86	12.51	92.59	91.07	728.57	247.66
S71	2.56	0.13	64.78	0.41	2.89	0.91	1.98	0.29	0.15	74.79	87.74	88.89	800.00	416.67
S72	3.00	0	87.88	0	0.92	0.58	0.57	0.80	0.65	26.46	75.49	77.78	233.33	61.73
S73	1.83	0.50	90.50	0.06	2.07	0.65	1.21	0.38	0.28	64.56	90.21	89.35	804.17	264.70
S74	3.00	0	59.84	1.08	1.79	0.77	0.90	0.44	0.33	66.18	99.19	93.33	466.67	122.22
S75	3.00	0	68.26	–	1.31	0.83	0.37	0.52	0.44	56.45	100.0	100.0	300.00	0
S76	2.56	0.50	89.30	0	2.15	0.83	1.14	0.46	0.28	67.16	92.72	90.00	540.00	214.81
S77	1.75	0.25	91.67	0.48	1.67	0.50	1.37	0.69	0.5	42.85	85.00	87.78	877.78	474.07
S78	3.00	0	77.68	0.41	1.81	0.78	0.85	0.44	0.32	67.29	98.85	95.00	475.00	113.89
S79	1.50	0.67	94.49	0.10	0.87	0.38	0.71	0.85	0.73	25.39	95.47	95.00	475.00	113.89
S80	2.67	0	70.00	0.37	2.20	0.95	1.18	0.33	0.21	71.52	90.32	91.67	458.33	291.67
S81	2.77	0	64.34	0.54	1.91	0.82	1.06	0.38	0.3	69.45	98.85	93.33	466.67	122.22
S82	1.52	0.25	95.71	0	0.64	0.27	0.75	0.90	0.82	17.74	98.73	80.00	400.00	377.78
S83	1.66	0.25	95.10	–	0.69	0.30	0.70	0.89	0.8	18.90	95.12	96.67	483.33	44.44
S84	3.00	0	88.60	0.06	0.81	0.51	0.46	0.83	0.7	29.05	95.99	94.44	283.33	61.73
S85	1.56	0.50	97.46	0.43	0.94	0.36	0.74	0.84	0.72	19.18	67.75	86.67	520.00	785.19

S86	3.00	0	61.73	–	1.91	0.95	0	1.00	1.00	0	0	0	0	38.58
S87	3.00	0	100.0	–	0		0	1.00	1.00	0	0	0	0	0
S88	1.78	0.17	82.18	0.32	2.10	0.75	1.18	0.52	0.32	65.44	96.29	88.10	616.67	479.97
S89	3.00	0	66.67	0	1.00	1.00	0.53	0.5	0.41	39.22	66.67	66.67	133.33	0
S90	2.14	0.50	76.67	0	1.38	0.87	0.64	0.57	0.40	56.86	95.24	88.89	266.67	246.91
S91	1.65	0.50	93.94	–	0.57	0.36	0.48	0.9	0.81	18.70	99.55	94.44	283.33	61.73
S92	2.19	0.13	79.78	0.69	2.05	0.65	1.35	0.4	0.32	66.25	97.11	89.35	804.17	372.73
S93	1.50	0	66.67	0.30	1	1	0.53	0.5	0.41	58.82	100.0	100	200	0
S94	1.50	–	100.0	–	0	–	0	1	1.00	0	0	0	0	0
S95	3.00	0	37.19	–	1.06	0.67	0.61	0.75	0.58	38.96	92.31	94.44	283.33	61.73
S96	2.50	0.50	66.67	–	1.58	1.00	0.67	0.33	0.30	70.18	100.0	100.0	300.00	0
S97	2.40	0	33.34	–	1.52	0.96	0.71	0.40	0.32	51.06	75.00	72.22	216.67	246.91
S98	3.00	0	40.69	0.20	1.08	0.42	0.93	0.81	0.67	28.52	85.99	86.67	520.00	266.67
S99	2.97	0.25	42.93	0.05	1.37	0.59	0.63	0.60	0.46	40.69	75.86	86.67	433.33	322.22

Note: FSI, Feeding Structure Index; ITI, Infaunal Trophic Index; P/A, Polychaetes/Amphipods Ratio; Δ, Taxonomic Diversity; Δ*, Taxonomic Distinctness; Δ+, Average Taxonomic Distinctness; TTD, Total Taxonomic Distinctness; STTD, Variation in Taxonomic Distinctness.

studies (e.g. Raposo, 1996; Costa *et al.* 1994), which consider the Mira estuary as representing what a pristine system should be.

4.3.3 The Mar Menor lagoon

The different environmental parameter values indicated that areas mostly affected by organic enrichment correspond to stations M2 and M6, where sediment organic-matter content reaches values higher than 5%. These stations also have in common the dominance of a polychaeta, being *Heteromastus filiformis* the most abundant species. The occurrence of lower values of Exergy and Specific Exergy, Taxonomic Diversity measures and *W*-Statistic should, therefore be expected, as well as higher values regarding *AMBI*, the Polychaete/Amphipod Index, FSI and the Infaunal Trophic Index. This was in fact confirmed by all indicators in station M6, but not in Station M2, where only the *W*-Statistic, Margalef Index TTD and *AMBI* indicated disturbance (Table 24).

Moreover, the Margalef Index and TTD are the only indicators capable to detect significant differences between organically enriched areas and non-enriched areas ($p < 0.05$) (Table 25). The *AMBI* values are similar in all sampling stations, indicating moderate disturbance in M2 and M6 and slight disturbance in the others. These results show that *AMBI* did not allow distinguishing between different disturbance intensities in this particular study area. Nevertheless, *AMBI* presents a positive response relative to the sediment organic-matter content, even if barely significant ($r = +0.41$; $p > 0.05$).

In its turn, the *W*-Statistic offers rather confusing results as in station M2, which presents sediment organic-matter content lower than in station M6, appearing, in this way, as the most polluted ($W = -0.3$).

In general terms, a similar pattern of variation was observed with regard to diversity measures and the Exergy Index, which showed positive and significant correlations ($p < 0.05$). On the other hand, these indicators were also negatively and significantly correlated ($p < 0.05$) with the sediment organic-matter content, as well as with other structuring factors within the system, such as salinity or, in the case of Margalef and Shannon–Wiener indices, with sediment particle size also. Specific Exergy showed a clear positive correlation with the presence of certain heavy metals such as Pb ($r = +0.89$; $p \leq 0.05$) and Zn ($r = +0.71$; $p \leq 0.05$), which is not what we would expect. For instance, station M2D, which presented the highest concentration of Pb and Zn, also exhibited a higher Specific-Exergy value.

Regarding Exergy values, the influence of biomass variations, which are related to numerical changes in the dominant populations under environmental stress, appear to be much more important than variations in the system's biomass quality (β factors). In the case of Specific Exergy, the influence of biomass variations is much less important, as changes in β factors related to the biomass quality play a major role. In this sense, the decline of taxonomic groups affected by toxic substances, as a function of different degrees of tolerance, should be clearly reflected in the values of Specific Exergy. Nevertheless, Molluscs, namely Bivalves, are known for their ability to bio-accumulate heavy metals, contrary to what happens,

Table 24: Indices' values estimated at the different sampling stations (M1–M10) in Mar Menor lagoon in A: July, B: November, C: February and D: May.

	AMBI	FSI	ITI	P/A	Shannon	Pielou	Margalef	Berger	Simpson	Δ	Δ*	Δ+	TTD	STTD	W-Statistic	Exergy	Sp-Ex
M1A	2.16	0.79	58.26	0.02	2.24	0.44	3.72	0.57	0.38	45.03	72.42	87.25	2966.67	403.52	0.44	2,885,503,836	14,9346
M1B	1.50	0.41	48.02	0.39	3.63	0.72	4.91	0.29	0.13	77.22	89.12	82.51	2722.92	532.79	0.17	183,671,203	155,184
M1C	1.50	0.47	45.59	0.03	2.19	0.43	3.87	0.61	0.40	43.71	72.74	88.15	2996.97	386.82	0.02	546,460,384	76,725
M1D	1.63	0.38	43.37	0.04	2.43	0.47	4.32	0.57	0.36	49.77	77.22	88.46	3273.15	370.33	0.26	192,624,681	70,963
M2D	0.10	0.20	33.34		2.75	0.77	2.13	0.28	0.18	56.45	68.89	85.10	1021.21	514.55	-0.3	15,762,446	603,402
M3D	3.57	0.09	67.46	0.89	2.06	0.49	3.20	0.68	0.47	42.48	79.71	77.67	1398.04	418.17	0.25	211,020	1592
M4D	0.39	0.50	86.50	0.04	2.14	0.55	2.02	0.39	0.28	63.16	87.15	88.25	1323.81	311.77		14,126,661,509	69,967
M6A	0	0	66.67				0	1.00	1.00	0	0	0	0	0	0.27	285,182	14,990
M5B	0.41	0.58	38.91	0.12	2.46	0.49	3.87	0.45	0.27	46.34	63.63	85.25	2727.96	430.36	4.58	899,957,796	109,861
M5C	0.44	0.67	37.81	0.14	2.55	0.56	3.26	0.42	0.25	46.10	61.49	82.19	1972.46	575.37	4.58	76,867,912	102,457
M6C	3.72	0.67	66.67	1.88	1.44	0.46	1.78	0.74	0.56	38.33	87.56	87.96	791.67	287.21	0.08	94,659	92,702
M5D	0.50	0.75	38.32	0.18	1.90	0.40	3.41	0.68	0.49	49.01	95.41	86.41	2246.67	407.63	-0.15	145,227,127	94,642
M6D	3.78	0	58.61	1.05	1.18	0.42	1.24	0.80	0.64	25.20	70.92	76.98	538.89	621.06	0.13	1,555,244	109,065
M7C	1.05	1.00	73.87	0	2.00	0.45	2.50	0.62	0.42	41.98	72.67	85.79	1801.67	421.20	-0.01	3,249,672,701	94,686
M8C	0.49	0.57	46.24	0.45	3.54	0.76	4.19	0.21	0.12	74.96	85.12	87.28	2181.94	407.59	0.11	70,381,987	67,160
M9	1.36	0.30	45.46		2.71	0.68	3.20	0.47	0.26	51.78	69.85	79.03	1264.44	340.26	-0.11	2,523,455	14,250
M11	1.69	0.60	72.0	1.26	3.75	0.75	5.12	0.26	0.12	73.76	83.93	79.44	2541.94	399.23	0.24	28,713,101	78,518
M10C	0.81	0	72.61	0	2.75	0.83	1.76	0.34	0.19	47.24	58.27	59.26	592.59	105.62	0.34	301,455,400	70,064

Note: FSI, Feeding Structure Index; ITI, Infaunal Trophic Index; P/A, Polychaete/Amphipod Ratio; Δ, Taxonomic Diversity; Δ*, Taxonomic Distinctness; Δ+, Average Taxonomic Distinctness; Δ+, Average Taxonomic Distinctness (presence/absence of species); TTD, Total Taxonomic Distinctness; STTD, Variation in Taxonomic Distinctness; Sp-Ex, Specific Exergy.

Table 25: Indices' values able to significantly discriminate (one-way ANOVA) organically enriched from non-organically enriched areas in the Mar Menor lagoon.

	Margalef	Total Taxonomic Distinctness
	Average	Average
Organically enriched areas	1.28	585.94
Non-organically enriched areas	3.92	1846.59

for instance, with Polychaetes, Crustaceans and Echinoderms. Since the β factor estimated for Molluscs is higher than that for the other groups mentioned (see Table 13), it is easy to understand why Specific-Exergy values were found to be higher in areas affected by heavy-metal pollution.

4.3.4 The Escombreras basin

In this study site, two criteria were established *a priori* to separate the stations into two groups, and to test the different ecological indicators' discriminatory capability. To begin with, sediment organic-matter content was taken into account and according to these criteria, stations E2, E7 and E8 were considered organically enriched. Secondly, an nMDS analysis was applied to data on taxa abundance to separate the stations into different groups (Figure 16).

In none of the cases did the indicators show capacity to discriminate between such groups. In general, results obtained with the different indicators were even contradictory. In fact, while diversity measures that take species abundance into account suggest a higher disturbance in station E8, diversity measures based on species richness indicate station E10 as the most polluted (Table 26).

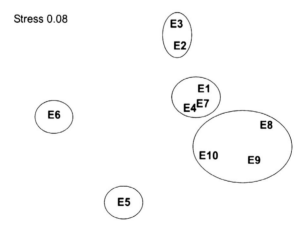

Figure 16 nMDS analysis. Two-dimensional plot of stations based on taxa abundance in the Escombreras basin.

Table 26: Values of ecological indicators estimated at the different sampling stations (E1–E10) at Escombreras basin.

	E1	E2	E3	E4	E5	E6	E7	E8	E9	E10
AMBI	3.47	2.33	1.59	3.04	1.00	–	2.45	0.65	2.14	1.29
FSI	0.33	0.60	0.50	0.50	0.67	–	0.38	0.33	0.33	0.67
ITI	70.00	52.23	49.02	57.15	83.34	–	50.38	89.72	81.77	90.48
Shannon–Wiener	2.21	3.02	1.33	2.28	1.79	–	3.45	1.24	1.41	1.38
Pielou	0.64	0.91	0.84	0.76	0.90	–	0.88	0.39	0.50	0.87
Margalef	2.17	2.65	0.71	1.87	1.67	–	3.68	1.76	1.24	1.03
Berger–Parker	0.52	0.22	0.58	0.42	0.50	–	0.20	0.78	0.50	0.57
Simpson	0.33	0.11	0.41	0.26	0.20	–	0.09	0.63	0.44	0.33
Δ	38.63	56.63	39.22	46.57	53.33	–	58.16	24.51	37.44	38.10
Δ^*	57.53	63.98	66.67	62.56	66.67	–	64.05	66.09	66.61	57.14
					200					
Δ^+	62.42	62.59	66.67	64.29	66.67	–	63.49	56.94	61.11	55.56
TTD	686.67	625.93	0	514.29	266.67	–	952.38	512.5	427.78	166.67
STTD	113.31	100.69	0	53.85	0	–	79.87	175.54	88.18	246.91
W-Statistic	–0.12	0.49	0.48	–0.02	0.50	–	0.56	–0.18	–0.08	0.12

Note: FSI, Feeding Structure Index; ITI, Infaunal Trophic Index; Δ, Taxonomic Diversity; Δ^*, Taxonomic Distinctness; Δ^+, Average Taxonomic Distinctness (presence/absence of species); TTD, Total Taxonomic Distinctness; STTD, Variation in Taxonomic Distinctness.

Other indicators, for instance the *W*-Statistic, mostly show a performance similar to diversity measures based on species abundance. On the other hand, *AMBI* indicates that station E1 is the most polluted, due to the dominance of *Polydora ciliata*, a polychaete belonging to ecological group IV, which is referred to by several authors (i.e. Pearson & Rosenberg, 1978; Gray, 1979; Villalba & Vietiez, 1985) as an indicator of organic pollution and oil pollution.

Finally, the performance of indicators based on ecological strategies was very unalike, not only when compared to each other, but also in comparison to indicators from other groups. In general, none of the indicators presents any significant correlation to physicochemical environmental parameters.

4.3.5 Cape Tiñoso

In the Cape Tiñoso case study, the populations' response to situations of disturbance versus non-disturbance was tested. Stations J4 and J5, the farthest from the influence of the cages, were considered as representing the reference situation. In addition, samples carried out in August 1996, prior to the placement of the floating cages, were considered as representing a pristine situation, in opposition to other sampling periods. Values estimated for the different indicators are given in Table 27.

In the first case, only *AMBI* was able to distinguish between reference stations and disturbed stations ($p < 0.05$), despite the fact that only polychaetes was taken into consideration. Nevertheless, such differentiation, even if statistically significant, proved to be irrelevant in terms of differentiating ecological status, as all the stations were generally identified as 'good' *sensu* Borja *et al.* (2000).

As for the comparison of samples from August 1996 with samples from later dates, none of the indices was able to illustrate the *a priori* assumed pristine situation, nor were they able to distinguish it from subsequent periods.

Once more significant correlations were found between diversity measures, Taxonomic Diversity and the *W*-Statistic, and also between the Margalef Index and TTD (Table 28).

On the other hand, and contrary to what should be expected, regarding the response of ecological indicators to environmental parameters, a positive correlation between several diversity measures, such as Taxonomic Diversity and TTD as well as chlorophyll *a* concentration in the water column was found (Table 29). These results suggest that the installation of floating cages allocated for red tuna fattening produced, at least during the first year and a half, a fairly small environmental impact, determining an intermediate disturbance situation, which in fact favoured a diversity increase.

The Average Taxonomic Distinctness was the exception, evincing a negative correlation (although not significant) with the concentration of chlorophyll *a* in the water column. This confirms the fact that the Average Taxonomic Distinctness response is monotonic, contrary to other diversity measures, as previously mentioned by Warwick & Clarke (1994).

Table 27: Indices' values estimated at the different sampling stations (J1–J10) in Cape Tiñoso.

		AMBI	ITI	Shannon	Pielou	Margalef	Berger	Simpson	Δ	Δ*	Δ+	TTD	STTD	W-Statistic
J1	A96	1.5	66.67	0.65	0.65	0.56	0.83	0.67	22.22	66.67	–	133.33	0	0.33
	N96	1.32	50.99	2.54	0.9	2.12	0.27	0.15	49.02	57.47	56.35	394.44	171.33	0.33
	F97	–	–	–	–	–	–	–	–	–	–	–	–	–
	J97	0.62	56.49	2.96	0.86	2.79	0.25	0.14	56.19	65.31	62.42	686.67	143.62	0.21
	N97	1.43	56.95	2.52	0.84	2.2	0.25	0.18	49.58	60.55	58.93	471.43	148.46	0.25
J2	A96	0.94	63.89	2.92	0.84	3.11	0.28	0.14	52.66	62.11	64.44	644.44	69.14	0.37
	N96	1.6	58.34	2.6	0.93	2.16	0.38	0.13	51.53	59.46	61.9	433.33	109.6	0.35
	F97	–	–	–	–	–	–	–	–	–	–	–	–	–
	J97	–	–	–	–	–	–	–	–	–	–	–	–	–
	N97	1.5	64.92	1.61	0.62	1.7	0.68	0.46	31.87	59.24	61.11	366.67	98.77	0.06
J3	A96	2.27	53.54	3.09	0.86	3.15	0.28	0.13	53.82	61.51	62.63	751.52	143.61	0.3
	N96	2.19	41.03	3.03	0.95	3.12	0.23	0.06	56.84	60.73	62.5	562.5	52.08	0.43
	F97	1.8	60	1.92	0.96	1.86	0.42	0.1	45	50	52.78	211.11	316.36	0.31
	J97	1.5	62.97	2.64	0.94	2.73	0.37	0.08	56.94	62.12	62.7	438.89	76.85	0.74
	N97	1.7	43.95	3.17	0.88	3.56	0.31	0.11	53.32	60.08	60.61	727.27	123.2	0.43
J4	A96	1.4	58.49	2.03	0.83	2.3	0.26	0.32	33.24	53.21	59.31	189.97	65.73	0.23
	N96	1.87	45.84	1.75	0.88	1.44	0.33	0.25	36.31	48.41	47.22	188.89	38.58	0.4
	F97	2.36	33.34	2.13	0.92	2.06	0.42	0.14	50	58.33	61.67	308.33	225	0.31
	J97	1.43	49.28	2.91	0.88	2.87	0.29	0.13	49.08	56.44	61.11	611.11	111.11	0.53
	N97	1.68	51.29	3.34	0.88	3.99	0.26	0.1	54.67	60.84	62.09	869.23	85.87	0.32
J5	A96	1.55	65.48	1.39	0.75	1.98	0.48	0.52	42.32	43.23	58.74	135.64	123.54	0.33
	N96	1.5	33.34	0.92	0.92	0.91	1	0.33	44.44	66.67	66.67	133.33	0	–
	F97	1.07	57.15	2.52	0.98	2.57	0.16	0.05	61.11	64.17	63.33	380	44.44	0.83
	J97	1.61	52.85	3.81	0.91	4.58	0.21	0.07	54.53	58.91	61.15	1039.58	151.37	0.48
	N97	1.12	51.01	0.81	0.81	0.72	0.5	0.5	33.33	66.67	66.67	133.33	0	0.46

(Continued)

Table 27: (Continued)

J6	A96	1.78	62.02	2.33	0.78	1.76	0.33	0.23	46.59	60.57	59.52	476.19	187.07	0.25
	N96	1.5	55.56	2.28	0.88	2.28	0.4	0.17	43.52	52.22	54.44	326.67	239.51	0.37
	F97	1.16	59.26	3.17	1	3.64	0.12	0	61.11	61.11	61.11	550	123.46	0.94
	J97	1.96	51.29	2.93	0.92	3.12	0.3	0.09	55.56	61.03	60.65	545.83	125.81	0.018
	N97	1.5	42.86	2.99	0.94	3.03	0.2	0.08	47.25	51.19	52.78	475	208.33	0.53
J7	A96	1.39	66.67	1.51	0.54	1.78	0.72	0.52	26.94	61.72	61.11	366.67	135.8	0.09
	N96	1.65	60	1.77	0.76	1.74	0.6	0.33	34.44	51.67	55	275	113.89	0.24
	F97	1.15	70.59	3.18	0.96	3.18	0.18	0.07	60.42	64.7	64.07	640.74	48.83	0.53
	J97	1.75	63.07	3.95	0.9	5.54	0.18	0.06	59.01	63.08	61.59	1293.33	103.83	0.6
	N97	1.68	62.97	2.28	0.88	2.28	0.5	0.17	46.76	56.11	58.89	353.33	106.17	0.23
J8	A96	1.67	63.89	1.3	0.5	1.33	0.74	0.57	17.53	40.49	61.11	366.67	172.84	-0.08
	N96	2	38.9	1.46	0.92	1.12	0.5	0.27	42.22	57.58	61.11	183.33	61.73	0.47
	F97	1.16	62.97	2.73	0.97	2.73	0.2	0.06	61.57	65.2	65.08	455.56	23.94	0.35
	J97	1.64	57.15	3.5	0.95	3.94	0.19	0.06	58.73	62.29	63.03	819.44	75.84	0.56
	N97	–	–	–	–	–	–	–	–	–	–	–	–	–
J9	A96	–	–	–	–	–	–	–	–	–	–	–	–	–
	N96	1.5	38.1	1.38	0.87	1.03	0.5	0.33	41.27	61.9	61.11	183.33	61.73	0.13
	F97	1.16	66.67	2.06	0.89	1.82	0.44	0.19	47.69	59.2	60	300	122.22	0.05
	J97	1.74	56.41	3.58	0.94	3.99	0.18	0.06	58.31	61.93	61.9	866.67	93.32	0.43
	N97	1.6	44.45	2.46	0.82	2.58	1	0.21	48.89	61.85	61.9	495.24	116.21	0.52
J10	A96	–	–	–	–	–	–	–	–	–	–	–	–	–
	N96	–	–	–	–	–	–	–	–	–	–	–	–	–
	F97	–	–	–	–	–	–	–	–	–	–	–	–	–
	J97	1.98	41.39	3.73	0.95	4.16	0.21	0.05	55.34	58.35	58.89	883.33	111.46	0.27
	N97	–	–	–	–	–	–	–	–	–	–	–	–	–

Note: ITI, Infaunal Trophic Index; Δ, Taxonomic Diversity; Δ^*, Taxonomic Distinctness; TTD, Total Taxonomic Distinctness; Δ^+, Average Taxonomic Distinctness; STTD, Variation in Taxonomic Distinctness; A96, August, 1996; N96, November, 1996; F97, February, 1997; J97, June, 1997; N97, November, 1997.

Table 28: Pearson correlations between the values of the different ecological indicators estimated based on data proceeding from sampling stations at Cape Tiñoso.

	Margalef	Δ⁺	Δ*	Δ	Pielou	STTD	TTD	Shannon	Simpson	Berger
Δ⁺	0.15									
Δ*	0.04	0.75**								
Δ	0.74*	0.42	0.16							
Pielou	0.46	0.24	-0.07	0.84*						
STTD	0.12	-0.56*	-0.56	-0.02	-0.03					
TTD	0.95**	0.19	0.15	0.61*	0.26	0.08				
Shannon	0.96**	0.13	-0.02	0.82**	0.55*	0.18	0.90**			
Simpson	-0.75*	-0.11	0.14	-0.94**	-0.86**	-0.21	-0.58*	-0.85**		
Berger	-0.71*	-0.32	0.24	-0.91**	-0.84**	0.23	-0.59*	-0.83**	0.98**	
W-Statistic	-0.55*	0.23	0.34	0.67*	0.76	0.34	0.45*	0.76	0.56*	0.57*

Note: Δ, Taxonomic Diversity; Δ*, Taxonomic Distinctness; Δ⁺, Average Taxonomic Distinctness; TTD, Total Taxonomic Distinctness; STTD, Variation in Taxonomic Distinctness (*): $p \leq 0.05$; (**): $p \leq 0.01$.

Table 29: Pearson correlations between the values of the different ecological indicators estimated based on data proceeding from sampling stations at Cape Tiñoso.

	Shannon	Simpson	Berger	Pielou	Δ	Margalef	TTD	Δ⁺
Chlorophyll *a*	0.46**	−0.39*	−0.41*	0.31*	0.37*	0.48**	0.38*	−0.23

Δ, Taxonomic Diversity; Δ^+, Average Taxonomic Distinctness; TTD, Total Taxonomic Distinctness (*): $p \leq 0.05$; (**): $p \leq 0.01$.

4.4 Was the ecological indicator's performance satisfactory in the case studies?

4.4.1 Indices based on indicator species

In general, *AMBI* worked reasonably well, being able to discriminate areas under pressure accounting for the benthic subtidal communities in the Mondego estuary, in the Mar Menor and in Cape Tiñoso. It was, however, inefficient in differentiating areas with clearly distinct eutrophication symptoms along a spatial gradient in the southern arm of the Mondego estuary (e.g. dominance of *Z. noltii* vs. *Ulva* sp. as main primary producers). In the case of the Mondego estuary, this may perhaps be explained if we accept that eutrophication effects, which are clearly visible at the primary producer's levels, are not strong enough to be detected by *AMBI* at other trophic levels.

In fact, although a number of species' composition shifts are already recognizable in qualitative terms, the structure of the benthic community in the three areas considered along the spatial gradient of eutrophication symptoms still exhibits, to a certain extent, a reasonably similar arrangement regarding the macrofaunal species (Marques *et al.*, 2003). In this case, *AMBI* values estimated in the Mondego estuary were similar at the three sampling areas due to the common dominance of *Hydrobia ulvae*, which belongs to ecological group III. Besides which all the other indicators were strongly affected by large abundances of *H. ulvae* and *Cerastoderma edule*, the dominant species. Nevertheless, such dominance does not have anything to do with pollution, being related rather to food resources availability (Pardal *et al.*, 2000). Despite these difficulties, with regard to other impact sources (e.g. outfalls, oil platforms), *AMBI* revealed to be efficient in detecting stress gradients (Borja *et al.*, 2003a). In fact, the application of *AMBI* in the Escombreras basin case studies and in the Cape Tiñoso case studies has lead to good results. For instance, in the case of Cape Tiñoso, even if accounting solely for Polychaetes, *AMBI* was the only indicator able to differentiate, although not very clearly, the control stations closer to the floating cages' area. *AMBI*'s good performance is also evident in the Mira estuary, where all sampling stations are regarded as being in good ecological status, which is fully consistent with other authors (e.g. Costa *et al.*, 1994; Raposo *et al.*, 1996).

On the whole, results lead us to think that *AMBI* is a good tool to detect pollution. However, some precautions, already described by Borja *et al.* (2003a), have to be taken to observe a correct application. It is assumed that *AMBI*'s robustness is reduced when only a very small number of taxa (1–3) and/or individuals are found in a sample. Moreover, to avoid ambiguous results, it is preferable to calculate *AMBI* values for each replicate separately, estimating the average value subsequently. When the percentage of unassigned taxa is elevated (> 20%) results must be evaluated with caution.

Some indicators, such as BENTIX, proposed by Simboura & Zenetos (2002), are based on *AMBI*. These authors modified *AMBI* by reducing the number of groups involved in the algorithm from five to three, to avoid errors when grouping species. However, in view of such modifications, BENTIX tends towards extreme values when evaluating a systems' ecological status. This type of response is due to only taking into account sensitive species (G.I.) and opportunist species of first and second order.

Other indicators, like the Norwegian Indicator Species Index (ISI), require a previous classification of sensitive values (based on the Hulbert Diversity Index) in the study area. For such a purpose, given a study area, a large number of samples are necessary, which is not easy in most cases. In the case of Norway, sensitive values for each species were determined after analysing 1080 samples from Norwegian fjords and coastal waters (1975–2001).

BENTIX's limited robustness and the difficulties arising in applying an index like ISI, make *AMBI* the most-useful index based on indicator species to establish the ecological status, at least for the present. Moreover, it has been tested in a large number of geographical areas and is supplied as user-friendly freely available software, including a continuously updated species list (approximately 3000 taxa presently), which makes it especially convenient.

4.4.2   Indices based on ecological strategies

The Polychaete/Amphipod Ratio was able to correctly illustrate the existence of an eutrophication gradient, using the Mondego estuary's southern arm intertidal communities. Nevertheless, in the other case studies, many samples did not allow to apply it simply due to the absence of amphipods. In such cases, the ratio would reflect an extremely polluted scenario, which, we knew for certain, was not the case. Although this indicator has been successfully used to detect the effects of organic and oil pollution on subtidal communities at the Bay of Morlaix (Mediterranean Sea) and at the Ría de Area and Betanzos (Atlantic Ocean), in our case studies it only worked well when applied to intertidal data. The Polychaete/Amphipod Ratio is probably influenced by a large spectrum of ecological factors, including some types of pollution. This means that this oversimplifying ratio is inadequate and difficult to relate to environmental quality.

Regarding indicators based on the trophic strategies (FSI and Infaunal Trophic Index), our results have shown their inefficacy as reliable tools in the detection of pollution. However, of the two, the FSI was more efficacious. In fact, while at least the FSI permitted discriminating between dredged and non-dredged areas on the subtidal communities of the Mondego estuary, the ITI was always inefficient in pointing out disturbance situations.

In addition, in the Mar Menor or in the Mondego estuary subtidal communities, contrary to what could be expected in accordance to Word (1990), ITI exhibited the highest values precisely in the less organically enriched areas. Actually, only in the case of the Mondego estuary's intertidal communities ITI values presented themselves as significantly correlated with the sediment organic-matter content. Currently, and precisely in this case, the higher sediment organic-matter content has a natural origin, the *Z. noltii* meadows primary production. Therefore, ITI does not appear able to differentiate between distinct situations along a gradient of eutrophication symptoms, which depend on the water column's nutrient concentration and water circulation (Marques *et al.*, 2003).

Apart from the bad results in the present case studies, there are other reasons not to recommend the use of ITI. One of the disadvantages of these type of indicators is the need to determine the organisms' diet, which can only be achieved through the study of stomach content, by laboratory experiments or through stable isotope analysis. As a rule, their actual diet is difficult to establish, and can vary between different populations from the same taxonomic entity.

Examples of such ambiguity took place when these indicators were applied to data from our study areas. *Nereis virens*, for instance, which is known as an omnivorous species along the European coast, becomes herbivorous in the North American coasts (Fauchald & Jumars, 1979). Also, *H. filiformis*, classified by Word (1990) as surface detritus feeder, is considered as subsurface deposit feeder by Brown (1985). What is more, while Word (1990) classifies most of the carnivore species in group II (surface detritus feeders), Codling & Ashley (1992) consider that they should belong to group III (surface deposit feeders), as most of them consume particles bigger than 50 micros in size. Another problem in determining the trophic category of many Polychaete species is their alternative feeding behaviour, which can occur under certain circumstances. For instance, through laboratory experiments, Buhr (1976) determined that the terebellid *Lanice conchylega*, considered as a detritivore, changes into a filter feeder when phytoplankton reaches a given concentration in the water column. Also, Taghon *et al.* (1980) observed that some species of the Spionidae family, usually taken for detritivores, could change into filterers, modifying the palps into a characteristic helicoidal shape. On the other hand, some species of the Sabellidae and Owenidae families can shift from filterers to detritivores. We can also determine that some omnivore and detritivore species become carnivores when they consume the remains of other animals (Dauer *et al.*, 1981; Maurer & Leathem, 1981). All these examples lead to doubts about the existence of a clear separation between different feeding strategies. That is why other characteristics, such as the degree of an individual's mobility and the morphology of the mouth parts must be included in the definition of Polychaete trophic categories (Gambi & Giangrande, 1985). Different combinations of such characteristics constitute what Fauchald & Jaumars (1979) named 'feeding guilds'.

On the benthic system studies, namely when identifying different types of impacts, authors like Maurer *et al.* (1981), Dauer (1984) and Pires & Múniz (1999) have tried to classify the different polychaetes species in feeding guilds with good results. The main problem in applying such a kind of classification is the determination of the possible combinations for each species. According to Dauer (1984), many families hold more than one combination depending on the type of feeding pattern they follow, their grade of mobility and the morphology of their mouth apparatus being every combination, therefore, monospecific. Very often, such a classification does not make much sense in practical terms.

Some of these questions, and the fact that the trophic group's classification in the case of ITI is not only based on where the food is captured, but on the size of the particle ingested, make the index even more difficult to apply in environmental studies.

4.4.3   Biodiversity as reflected in diversity measures

Regarding diversity measures, the Margalef Index presented the best perform-ance, despite its relative simplicity as compared to other indices, namely account-ing for species richness and abundance of individuals. In fact, it successfully differentiated distinct eutrophication levels in the Mondego estuary's southern arm intertidal communities, and was also effective in detecting organic enrich-ment situations in the Mar Menor lagoon. As for the Shannon–Wiener, the Simpson and the Berger–Parker indices, they appear to be too influenced by the dominance of certain species (e.g. *H. ulvae* in the Mondego estuary or *Bittium* sp. in the Mar Menor), whose abundance has no relation with any type of distur-bance, being favoured rather by abundant food resources.

Of all the indicators based on Taxonomic Distinctness, only TTD was able to correctly distinguish between different scenarios along the gradient of eutrophi-cation symptoms in the southern arm of the Mondego estuary. Moreover, together with *AMBI* and the Margalef Index, TTD showed that it was capable of discrimi-nating between more and less organically enriched areas in the Mar Menor. In all our case studies, TTD appears significantly correlated to the Margalef Index, and in the case of the intertidal communities in the Mondego estuary, it exhibits itself as the most sensitive of the two. Nonetheless, Warwick & Clarke (1998) deem the use of that measure as not recommendable due to the fact that, in general, TTD tends to track species richness rather closely, and it is only useful for tightly controlled designs in which effort is identical for the samples being compared, or sampling is sufficiently exhaustive for the asymptote of the species–area curve to be reached.

Although in theory the other Taxonomic Distinctness measures cover many of the features (e.g. independency on simple size/effort or monotonic response to environmental degradation) required to be a good diversity indicator, in view of our results, the other measures proposed by Warwick & Clarke (1995; 1998) did not show any advantage as compared to other diversity indices. An exception was the Cape Tiñoso case study, where we observed a situation of intermediate organic enrichment, susceptible in some cases of favouring an increase in diversity. Here, contrary to what happens with other diversity indices, Average Taxonic Distinctness is not positively correlated with the concentration of chlorophyll *a* in the water column, exhibiting, in fact, a monotonic answer to stress.

The fact that, in general, Taxonomic Distinctness measures do not appear to be more sensitive to environmental stress, as compared to other diversity indices, has also been observed by Somerfield *et al.* (1997) in studies on the North Sea impact of oil fields and Hall & Greenstreet (1998), studying fish communities, found that Taxonomic Distinctness measures showed identical trends to conventional diversity indices. Yet, Somerfield *et al.* (2003) have proposed Average Taxonomic Distinctness to be used as a tool in the classification of the ecological status when implementing the WFD. This recommendation is due to an apparent advantage, which is the fact that it includes a master list of *taxa* corresponding to what is assumed to represent reference conditions. Moreover, the software includes a statistical framework from which to measure the departure from what is expected (the reference condition).

In spite of such advantages, a study by Prior *et al.* (2004) suggests that before considering Taxonomic Distinctness measures as applicable in the WFD scope, some modifications have to be introduced. In fact, investigations have shown that Taxonomic Distinctness is sensitive to the frequency of occurrence of taxa across each sample. This is contrary to the null hypothesis upon which Average Taxonomic Distinctness is currently based, that takes into account the natural spatial variation caused by reproductive strategies within benthic communities. That is why Prior *et al.* (2004) underline the need for frequency distribution to be well studied for high ecological status, to set a strong reference from which to measure departures.

It is interesting to observe how the two tested indices based on specific richness (Margalef index and TTD) were the most successful measures in differentiating the diverse grades of pollution, leading us to think that the increment or decrement in the number of species is one of the best disturbance indicators, and therefore, essential when it comes to differentiating ecological status.

The Northeast Atlantic Geographical Intercalibration Group Benthic Expert (NEAGIG, 2004) considered that the selected metrics to be used in the WFD context need to distinguish clearly the good/moderate boundary. Obviously, those two measurements alone are not able to work out such a distinction, as they will always need a previous knowledge on the number of species (reference situation) of the studied site. In that sense, few are the indices capable of establishing the different ecological status (high, good, moderate, poor and bad).

The other inconvenience of species richness is that, contrary to Taxonomic Distinctness, it may be more sensitive to underlying variation in natural environmental factors, thus generating confounding effects if one is interested in the influence of anthropogenic perturbations (Warwick & Clarke, 1998; Leonard *et al.*, 2006). Indeed, the fact that the Average Taxonomic Distinctness sustained high values in the Mondego stations with few species but low levels of organic matter and therefore not organically polluted, showed the ability of this index to detect impacts despite possible natural environmental disturbances, as for example salinity fluctuations, in an estuary or coastal lagoon. Those salinity fluctuations in the Mondego estuary do not seem to be the determining factor influencing species' richness. The previous-mentioned measure can be affected by a number of factors, as for example the marine water and freshwater inputs that make the colonization and settlement possibilities of certain species difficult. Nevertheless, in the Mar Menor the salinity was a very influential parameter once it is correlated with the lagoon confinement or isolation degree (Gamito *et al.*, 2005; Pérez-Ruzafa & Marcos, 1992; Pérez-Ruzafa *et al.*, 2005).

Studies such as the one performed by Heino *et al.* (2005) showed that Taxonomic Distinctness also varies along natural gradients and it is unlikely that a site can be determined to be degraded or not degraded based only on this measure. On the other hand, although Average Taxonomic Distinctness has the ability to discriminate properly between polluted and non-polluted areas in those with low number of species (as is the case of the subtidal communities in the Mondego), the results of this study demonstrated that its power of discrimination decreases when the species number increases (see confidence limits in the funnel graphic representation, Figure 12), which leads us to believe that the index is not able to show correlations with pollution in areas where richness depends on other factors.

4.4.4   Indicators based on species biomass and abundance

In most of our case studies, the *W*-Statistic appears significantly correlated with the Shannon–Wiener, Pielou, Berger–Parker and Simpson indices, but it presents a clear comparative advantage: its application does not depend on previously known reference values.

Nevertheless, the dominance of few species with small-sized individuals, although characteristic of polluted environments, may occur in non-polluted environments, not at all unusual for instance in the Mondego's estuarine benthic community, which may lead to erroneous ecological status assessment. This problem has in fact been perceived in several case studies (Ibanez & Dauvin, 1988; Beukema, 1988; Weston, 1990; Craeymeersch, 1991), and is the reason why the *W*-Statistic was not very successful in detecting organic pollution in the Mar Menor lagoon or at the Escombreras basin. A possible explanation for this is the fact that the *W*-Statistic was entirely developed to assess the impact of organic pollution, and in these two study areas, although sediment organic enrichment is a concern, there are also other kinds of pollution (e.g. heavy metals), and different types of environmental stresses.

4.4.5   Thermodynamically oriented indicators: Exergy-based indices

As a whole, our results suggest that the Exergy Index is able to capture useful information about the state of the community. In fact, more than a simple description of the system's environmental state, the spatial and temporal variations of the

Exergy Index may provide a much better understanding of the system's development in the scope of a broader theoretical framework.

However, at the present stage, through simple snapshots, the Exergy Index and Specific Exergy can hardly provide a clear discrimination between disturbed (i.e. polluted) and non-disturbed situations. For instance, in the case of the Mar Menor marine lagoon, despite responding to sediment organic enrichment, both the Exergy Index and the Specific Exergy were unable to distinguish between areas affected by organic pollution and areas that are not. Nevertheless, the Exergy Index worked relatively well regarding the Mondego estuary's intertidal communities, being able to distinguish between different areas along a gradient of eutrophication symptoms. These differences in efficiency might be owing to the fact that in the Mar Menor lagoon the effects of organic pollution are, to a certain extent, covered up by other system-structuring factors, while in the Mondego estuary the southern arm eutrophication was undoubtedly the major driving force behind the ongoing changes.

Finally, in the case of the Mar Menor, it is interesting to note that Specific Exergy appears positively correlated to heavy-metal contamination (such as lead and zinc), whereas the Exergy Index does not, which is basically on account of their different responses to biomass variations in the community. In fact, the influence of such variations on Specific-Exergy values is far less important, because weighting factors expressing the quality of biomass play a major role in estimations.

4.4.6 Integrative indices: B-IBI

B-IBI was only applied on the subtidal communities of the Mondego estuary and did not prove sensitive enough to distinguish between different *a priori* well-known zones.

Although one of the B-IBI issues is the balance (%) between species sensitive and tolerant to pollution, which should work fairly well, it also accounts for the percentage of trophic groups and diversity, measured by the Shannon–Wiener Index that, as mentioned before, did not work in distinguishing different levels of eutrophication in the Mondego estuary. Our results appear to indicate that B-IBI is system specific, and because of this, its effectiveness depends on the geographical area where it is applied.

In fact, while the index was satisfactory at the Chesapeake Bay and New York–New Jersey harbour areas, it underwent adaptations to allow its correct application in other areas. For instance, Van Dolah *et al.* (1999) considered four metrics to use B-IBI in Carolina: (a) mean abundance, (b) mean number of taxa, (c) 100 minus percent abundance of the top two numerical dominants and (d) percent abundance of pollution sensitive taxa, without diversity values. In the Mondego estuary case study, it would possibly have been better to reflect on other types of issues, such as the percentage of abundance of pollution sensitive and pollution tolerant species, and diversity measured as species richness, which proved to work well in this system, instead of considering the proportional abundance of individuals as well. But in that case we would be applying a different index, than the one proposed by Weisberg *et al.* (1997).

Surely, the major inconvenience of an index such as B-IBI is the unavoidable need to readapt it to different geographical areas. The basic steps to develop these types of indices are: (a) defining major habitat types based on classification analysis of species composition and evaluation of the physical characteristics of the resulting site groups, (b) selecting a development data set representative of degraded and non-degraded reference sites in each major habitat type, comparing various benthic attributes between them and (c) establishing a scoring criteria. Obviously, this implies previous knowledge on the study areas, and the availability of a large database (which in most cases does not exist), to validate the measures, and such constraints lead us to discourage the generalized application of B-IBI.

5 Combining indicators to characterise a system's ecological quality status

Ecological indicators, based on monitoring activities, are commonly used to assess quality status and provide support on the management of natural resources. Nevertheless, due to the natural complexity of ecological systems, the selection of appropriate indicators to deal with such questions is quite difficult. In fact, to increase the reliability of assessment results, it is usually indispensable to use a set of indicators representing the structure, function and composition of such ecological systems.

Our results illustrate that none of the effects of disturbance measures presently available can be considered perfect when used alone. Consequently, in our opinion, one should always consider a suitable combination of several indicators to make up for each of their individual limitations. This may result in a good toolset to assess the ecological quality status (EQS) of an ecosystem. Let us exemplify this in more detail.

AMBI, the Margalef Index and Total Taxonomic Distinctness, for instance, appeared as the most sensitive indicators in discriminating disturbance situations in our case studies. They must therefore be considered as good tools in any multimetric approach aiming at establishing ecological levels. However, *AMBI*, Margalef Index and the Total Taxonomic Distinctness require reference values. This makes their integration in a combined approach difficult. In fact, establishing a correspondence between indicator values and ecological status turns out to be necessary.

In our study on the intertidal communities of the Mondego estuary, Total Taxonomic Distinctness was shown to be more sensitive to environmental changes than the Margalef Index. Nevertheless, it was deemed preferable not to integrate it in methodologies involving the combination of different indicators because, according to Clarke & Warwick (1999), it is of limited applicability. These authors consider that Total Taxonomic Distinctness tends to track species richness rather closely, and will be useful only for tightly controlled designs, where effort is identical for the samples under comparison, or when sampling is sufficiently thorough to guarantee that the asymptote's species–area curve is reached.

On the other hand, even though in our Mondego estuary case study the Shannon–Wiener Index performed less well than the Margalef Index, it is still our opinion that the Shannon–Wiener Index has to be taken into account because diversity measures must address aspects (i.e. proportional abundance of individuals) other than species richness. Moreover, even if the application of the

Shannon–Wiener Index requires reference values, the fact that it is probably the most used diversity measure in environmental studies makes it relatively easy to establish relationship between values obtained and different levels of ecological status. These are few examples of the difficulties encountered, but even then the complementary use of various indicators or methods, based on different ecological principles, is highly recommended in determining an ecosystems' environmental quality status.

The implementation of the European Water Framework Directive (WFD, 2000/60/CE), as well as the recent European Marine Strategy, is certainly an important field of application. In fact, the approach to water issues has changed significantly since the WFD became effective. The existence of the main goal of achieving 'good ecological status' in all water bodies (i.e. surface and ground-water) within a given period has triggered numerous proposals of multimetric ecological quality assessment tools by the European scientific community (e.g. Borja et al., 2004; Rosenberg et al., 2004; Bald et al., 2005; Simboura et al., 2005; Loureiro et al., 2006; Devlin et al., 2007; Patrício et al., 2007; Blanchet et al., 2008; Neto et al., submitted; Teixeira et al., submitted).

In the scope of the WFD, a water body ecological quality status (EQS) is primarily determined according to its biological quality elements and at a later stage by using a range of hydromorphological and physicochemical quality elements. For transitional waters, the biological quality elements to be assessed as mentioned by the WFD (2000/60/CE) are phytoplankton (composition, abundance and biomass), other aquatic flora (composition and abundance), benthic invertebrate fauna (composition and abundance) and fish fauna (composition and abundance). In the case of coastal waters (CW), the biological elements include phytoplankton, other aquatic flora (macroalgae and angiosperms) and benthic invertebrate fauna, considering the same measurable factors mentioned above. For practical reasons, fish fauna is not taken into consideration when assessing the EQS of coastal water bodies.

The final classification of a water body will reflect to what extent its EQS deviates from a reference condition (i.e. the expected values of the measured elements for that type of system in a non-disturbed situation (WFD, Annex V)). Therefore, reference conditions need to be defined for all the typologies presented in the scope of the WFD (Vincent et al., 2002); they must also summarise the range of possibilities and values for the biological quality elements over periods of time and across the geographical extent of the type (EC, 2003). Although reference conditions represent a part of Nature's continuum (e.g. the European Sea) and must reflect natural variability, the different typologies considered artificially compartmentalise this continuum into a number of physical types. Consequently, the reference conditions for a specific water body type must describe all the possible natural variations within that type (Figure 17) (EC, 2003). This means that to proceed to a correct ecological evaluation, the same reference conditions might not be adequate for the entire system (Borja et al., 2008). Another example may be taken from transitional waters, such

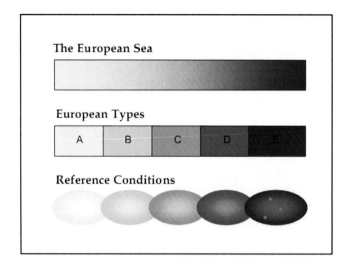

Figure 17. The figure shows the relationship between the seas in Europe (the European Sea) and the typology and type-specific reference conditions. It
also shows how sites within a type may be used to establish the natural
variability within the type (EC, 2003).

as estuaries, where due to high variability of physical conditions and strong
gradients from the mouth to the head, different conditions are expected to exist
throughout the system (Best *et al.*, 2007; Muxika *et al.*, 2007; Teixeira *et al.*,
2008a). In this case, for an efficient assessment, we need to define stretches,
distinct habitats for which appropriate reference conditions must be determined
(Teixeira *et al.*, 2008b). Besides the hydromorphological and physicochemical
aspects, which are biologically determinant, other characteristics such as pressure and state indicators may be helpful in delimiting specific stretches within
a system (Ferreira *et al.*, 2006).

Reference conditions should be derived in view of distinguishing among very
minor, slight and moderate disturbances. 'Very minor' disturbance could be
defined as barely detectable in the sense that the disturbance is more likely to be
anthropogenic than not. Slight disturbance could be defined as anthropogenic at
a prescribed level of confidence.

Although this process is still under implementation, it is our belief that the
right use of such a multimetric approach can be of great utility in establishing EQS in environmental studies. Hopefully, in the near future, studies of a
more empirical nature will provide further examples of the effectiveness of this
approach.

In this chapter, we illustrate the application of three multimetric approaches
for EQS assessment, in the scope of WFD, in a transitional (Mondego estuary)
and in a coastal system (Buarcos bay)

5.1   Ecological quality status assessment – multimetric approaches within the WFD

5.1.1   Transitional waters: The Mondego estuary case study

The Mondego River basin has an area of approximately 6670 km^2, including a large alluvial plain consisting of good-quality agricultural land. The construction of harbour facilities in the river's estuary introduced changes into the system since the 1930s. However, it was from the 1960s that the Mondego's catchment area underwent large-scale morphological modifications. These involved the construction of stone walls and water reservoirs to regulate the river's water flow, improve the uses of water resources in industry and agriculture, produce electric power and expand the harbour facilities. All this had a strong anthropogenic impact on the system, modifying the riverbed's topography and changing hydrodynamics.

The entire catchment area of the Mondego River, particularly the direct run-off from the 15,000 ha of cultivated land in the lower river valley (mainly rice fields), presently contributes with a considerable loading of nutrients and several chemicals into the estuary (western coast of Portugal; 40°08′N, 8°50′W), which constitute a relatively small (860 ha) warm-temperate polyhaline intertidal system. In addition, the Mondego estuary supports industrial activities, salt works and aquaculture farms, being the location of the commercial and fishing harbours of Figueira da Foz, and also a centre of seasonal tourism activity.

The estuary is 21-km long and its terminal part, 7-km long and 2–3-km across at its widest section, consists of two arms, north and south, separated by the Murraceira Island. The northern arm, where the commercial harbour is located, is deeper (5–10 m during high tide) and constitutes the main shipping channel. The southern arm is shallower (2–4 m during high tide) and is characterised by large areas of intertidal mudflats (almost 75% of the area) exposed during low tide.

In the early 1990s, the south arm was heavily silted in the upstream areas, causing the river discharge (from 27 m^3 s^{-1} in dry years to 140 m^3 s^{-1} in wet years; mean annual average 79 m^3 s^{-1}) to flow essentially through the northern arm. Consequently, water circulation in the south arm became mostly dependent on tides and the reduced freshwater input from the Pranto River, a small tributary artificially controlled by a sluice (Marques *et al.*, 2003). The tidal range varies between 0.35 and 3.3 m depending on site and tide coefficient, while water residence time varies between 2 days, in the northern arm (Marques *et al.*, 2007), and 5 days, in the southern arm, with a very small water discharge from the Pranto sluice (Neto *et al.*, 2008).

During the 1990s, two major hydraulic changes occurred in the estuary:

(a) From 1990 to 1994, the North Channel sunk 1.5–2 m and the banks filled, narrowing the width of the river bed by about 75% (Figure 18). This reduction allowed the water to flow faster, aiding in preventing the silting that would otherwise normally occur in these parts of the river.

Figure 18. The figure shows final interventions at the north arm of the Mondego estuary. The situation in 1994 is depicted as follows: area between the commercial harbour (at the mouth) and the point of separation of the two arms. Details of the left margin's land filling process are also provided.

As a consequence, the communication between the two arms of the estuary was totally interrupted in the upstream area due to the completion of stone walls in the north arm banks (Figure 19). In fact, four water pipes with a Ø 0.4 m section were kept to ensure minimum communication, but these rapidly became totally silted up and the water stopped running through them.

Figure 19. The figure shows the communication between the two arms of the Mondego estuary after the northern arm interventions in 1994. Details of the water pipes are also presented.

(b) During the period between 1997 and 1998, the following experimental interventions were carried out to decrease eutrophication symptoms (green macroalgal blooms occurred since the early 1990s) and test ways of ameliorating the system's condition:

- The freshwater discharge of the Pranto River sluice into the south arm was reduced to a minimum to decrease nutrient loading. It was instead diverted to the northern arm by another sluice located further upstream.
- The communication between the northern and south arms was re-established, although only for a very limited period (only 1.5–2 hours before and after each high tide peak and through a section of only 1 m²), to improve water circulation (Figure 20) (Neto, 2004; Lillebø et al., 2005; Neto et al., 2008).

Figure 20. The figure shows details of the experimental re-establishment of the communication between the two arms of the estuary late in 1997.

From the mid-1990s to 2006, the main pressures affecting the Mondego's northern arm resulted mainly from harbour facilities and consequent dredging activities, causing physical disturbance of the bottoms.

A long-term study of the Mondego estuary, pursued since the mid-1980s, allowed to follow the system's response, not only in terms of changes in physical conditions but also with regard to the occurrence of extreme climatic events. Important quality elements, such as water quality, hydraulics and sediment dynamics, benthic intertidal and subtidal communities, changes in *Zostera noltii* beds and green macroalgae distribution, have been monitored since then. The main alterations (Figure 21) can be roughly summarised as follows:

(a) In spite of morphological changes artificially introduced in the natural river course since the 1960s, eutrophication symptoms were not noticeable in the estuary prior to the early 1990s.

(b) Following the interruption of the upstream communication between the two arms, especially from 1991 to 1997, the ecological conditions in the south arm showed a rapid deterioration. The combined effect of increased water residence time and nutrient concentrations became the major driving force behind the emergence of clear eutrophication symptoms. Seasonal blooms of *Ulva* spp. were observed concomitantly with a severe reduction of the area occupied by *Z. noltii* beds, as a function of spatial competition with macroalgae (Marques *et al.*, 2003).

(c) The shift in benthic primary producers affected the structure and functioning of the biological communities, and through time, such modifications started inducing the emergence of a new selected trophic structure, which has been

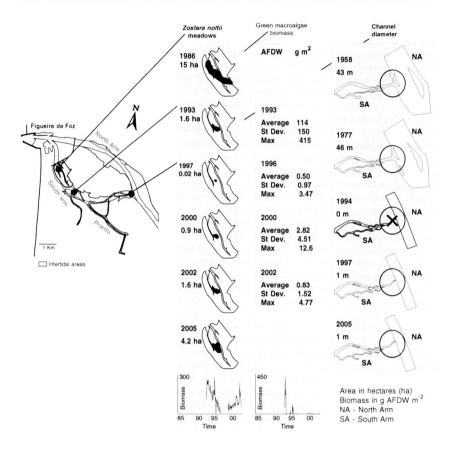

Figure 21. The figure shows the temporal changes of *Zostera noltii* meadow area
(ha) and biomass (g AFWD m^{-2}), green macroalgae biomass (g AFWD
m^{-2}) and the connection width between the two estuarine arms. Black
areas correspond to *Z. noltii* meadows, eutrophic areas and connection
between the two arms (from Patrício *et al.*, submitted).

analysed abundantly in literature (e.g. Cabral *et al.*, 1999; Dolbeth *et al.*, 2003;
Cardoso *et al.*, 2004a, 2004b; Lopes *et al.*, 2005, 2006; Marques *et al.*, 1997,
2003; Martins *et al.*, 2005, 2007; Patrício *et al.*, 2004; Patrício & Marques,
2006).

(d) From 1998, following the reduction of freshwater discharge proceeding from
the Pranto River sluice and the limited re-establishment of the upstream communication between the two arms (1997/98), this trend appeared to reverse to
a certain extent. This is suggested by the partial recovery of the area occupied
by *Z. noltii*, previously the richest habitat in terms of productivity and biodiversity (Marques *et al.*, 1993), and the cessation of green *Ulva* spp. blooms
(Lillebø *et al.*, 2007; Verdelhos *et al.*, 2005).

The full re-establishment of the communication between the two arms appeared, therefore, as a suitable way of improving the system's ecological quality (Marques *et al.*, 2005), an intervention that was undertaken during the spring of 2006.

Based on the Mondego estuary case study, our intention was to illustrate the way these multimetric approaches, designed to assess ecological status, track the history of human-induced changes, such as the one described above, and to consolidate the impact they have on the ecological assessment's practical application.

5.1.1.1   Opportunistic macroalgae EQS assessment

The methodology underlying the approach taken for the development of a tool directed to monitor mats of various bloom-forming macroalgae on intertidal sedimentary shores was described by Scanlan *et al.* (2007), and in this case, the main pressure considered was also eutrophication.

Scanlan *et al.* (2007) do not recommend any particular survey method to monitor macroalgal blooms. Instead, they state that the pros and cons of survey methods, as well as costs, should be carefully considered. They enumerate a list of possible methodologies, from conventional techniques to remote sensing methods, to accomplish the objective.

Regarding the sampling period, Scanlan *et al.* (2007) claim that, depending on local patterns, it may be necessary to monitor macroalgal blooms during both spring and summer. In the United Kingdom and Republic of Ireland, the biomass peak is mostly found in late summer. Therefore, the above-mentioned authors recommended that monitoring should take place during this period. In Portugal, results from the Mondego estuary suggest that sampling should be carried out from April to June (Patrício *et al.*, 2007). Nevertheless, although they consider that, ideally, mats might be monitored throughout their growing season, they also state that this requirement would be highly resource intensive and could underestimate impacts at peak times.

5.1.1.1.1   Basic parameters   A selection of tools was chosen based on the parameters that would describe or indicate the response of macroalgae to disturbance. It is not expected to use any one of these single metrics in isolation to understand ecological patterns or to derive a classification.

Four basic parameters have been proposed:

A   Total available intertidal area for opportunistic macroalgal growth (ha)
B   Areal coverage (ha)
C   Percentage (%) cover
D   Biomass (g WW m^{-2})

5.1.1.1.2   Opportunistic macroalgae assessment method – Option 1   Option 1 is the approach proposed by Scanlan *et al.* (2007). It combines, in the first step, the percentage cover with biomass to obtain a classification (see Table 5, Chapter 2, Section 2.3.1.). For further details regarding boundary conditions, please refer to the mentioned paper.

Secondly, to account for overall water body size, the authors proposed that areal coverage (in hectares) should lower the class of a water body, as derived from Table 6 (Chapter 2, Section 2.3.1.), by one or more classes depending on the total area of algal mats (Scanlan *et al.*, 2007).

5.1.1.1.3   Opportunistic   macroalgae   assessment   method   –   Option   2
An alternative less-costly approach – Option 2, suggested by Patrício *et al.* (2007) – is to combine only percentage cover with areal coverage (in hectares). Table 7 provides an initial classification status, but does not account for the overall size of the patch within the water body. Table 6, on the other hand, allows taking into account situations where there is a large algal bloom but the percentage cover is relatively small due to the large size of the water body. Consequently, areas where there is an extensive bloom of algae cannot achieve high status.

5.1.1.1.4   Data set   The intertidal zone of the south arm of the Mondego estuary, consisting of soft sediments (mud, muddy sand and sand), was chosen as the study site. The sampling campaigns were included in a wider monitoring survey and took place monthly, from January 1993 to September 2000.

5.1.1.1.4.1   Total intertidal area   Based on aerial photographs of the entire southern-arm area and using GIS technology, the intertidal areas suitable for opportunistic macroalgal growth were estimated on 175 ha.

5.1.1.1.4.2   Macroalgal cover   The area of macroalgae-covered sediments in most of the intertidal areas of the south arm was plotted on maps drawn from aerial photographs. To increase plotting accuracy, the maps included all the identifiable landmarks in the surroundings, such as channels, patches of plants, electricity poles and nearby buildings. The area of macroalgal cover was then estimated from the maps using a grid where each square corresponded to 100 m^2. Additionally, to each defined area one of three percentage of cover categories was attributed : (1) 0% (no algae); (2) 50% (patchy distribution of algae covering around 50% of the area) or (3) 100% (algae covering the entire area without bare sediment). Estimations were confirmed through field observations. Macroalgal cover of the entire system was then calculated according to Wither (2003) (for more details see Section 5.1.2.2.).

5.1.1.1.4.3   Biomass   During low tide, a manual corer with a section of 141 cm^2, introduced in the sediment to a depth of 20 cm (five replicates of approximately 3 L) was used to collect the opportunistic macroalgae occurring on the intertidal areas' soft substrates. To classify the whole water body, biomass figures (in g WW m^{-2}) are expressed as mean values for the entire available intertidal area.

5.1.1.1.5   EQS results   Both options captured the inter-annual variations regarding macroalgal dynamics in the study area, presenting worse classifications in 1993 and 1995, which is in accordance with our understanding of the system's

Table 30. EQS for the south arm of the Mondego estuary, April–June from 1993 to 2000, based on Option 1: % cover, biomass production of opportunistic macroalgae and areal coverage, and Option 2: % cover and areal coverage of opportunistic macroalgae.

Year	Option 1				Option 2		
	% cover	Biomass (WW g m^{-2})	Areal coverage (ha)	EQS	% cover	Areal coverage (ha)	EQS
1993	32.42	294	57	M	32.42	57	P
1994	13.34	7	23	G	13.34	23	G
1995	27.39	74	48	G/M	27.39	48	P
1996	4.32	1	8	H	4.32	8	H
1997	10.01	9	18	G	10.01	18	G
1998	6.93	6	12	G	6.93	12	G
1999	2.52	0	4	H	2.52	4	H
2000	3.09	0	5	H	3.09	5	H

Note: Ecological quality status (EQS): H, High; G, Good; M, Moderate and P, Poor.
Total available intertidal area for opportunistic macroalgae growth in the south arm = 175 ha.

evolution. After the implementation of preliminary mitigation measures to decrease the eutrophication symptoms felt in the system, the area exhibited a positive evolution with the reduction of macroalgal blooms and the recovery of *Z. noltii* meadows. Both methodological options accurately captured these dynamics (Table 30).

In 1994, a classification of Good was produced by both options. At first, this result might seem contradictory with the results obtained in 1993 and 1995. However, it is well known that in this system macroalgal blooms may not occur in rainy years, due to long intervals of low salinity coupled with stronger currents caused by the discharge from the Pranto River (Martins *et al.*, 1999, 2001, 2007). In reality, the accumulated precipitation (from September to June) in 1994 (749 mm) was higher than that in 1993 and 1995 (565 and 615 mm, respectively) (Patrício *et al.*, 2007), which could explain the inexistence of macroalgal blooms. Although the year 1999 presented low accumulated precipitation (486 mm), this occurred after the implementation of experimental mitigation measures that, among other effects, increased the system's hydrodynamics, avoiding the development of macroalgal blooms.

The year 1997 exhibited a Good EQS, despite the fact that based on our knowledge of the system the condition was worse than in 1996 or 1998–2000. None of the methodological options was able to capture this trend regarding the system's evolution.

Results suggest that both options captured the essence of inter-annual variations of macroalgal dynamics in the south arm of the Mondego estuary, providing worse classifications in 1993 and 1995. According to our understanding of the system, these results reflect the combined effect of increased water residence time and high nutrient concentration, which caused the emergence of clear eutrophication symptoms since the early 1990s (Marques et al., 2003). In 1998, after the implementation of preliminary mitigation measures, which improved water circulation in the south arm of the estuary, the area began to show signs of recovery, such as the cessation of extensive green macroalgae blooms, among others (see Figure 21). Again, both assessment methods correctly captured this positive evolution of the system.

Moreover, in all situations, Options 1 and 2 agreed on the classification regarding the distinction of Good and High EQS, suggesting that, at least in this estuary, the threshold values proposed by Scanlan et al. (2007) seem to be adequate for these two classes.

Nevertheless, Option 1 at times upgraded one class status below Good when compared with Option 2. In such cases, results obtained with Option 2 were more in accordance with our empirical perception of the real condition of the system. Therefore, the match between the EQS resultant from each of the two methodologies would increase by avoiding the hybrid classifications provided by Option 1 (e.g. Good/Moderate or Moderate/Poor). In other words, if the Good/Moderate classification given by Option 1 were changed to Moderate and the Moderate/Poor changed to Poor, by slightly adjusting the threshold values, both options would give similar results. Taking into account the WFD definitions (WFD, 2000/60/CE), we suggest the use of the scale proposed by Option 2, not only because it avoids imprecise classifications, but also because it accomplishes the precaution principle and appears to be more in accordance with our empirical understanding of the system's environmental condition throughout the study period.

5.1.1.2   Benthic macroinvertebrates EQS assessment

In the scope of WFD, macrobenthos is one of the biological quality elements to be taken into account for the quality assessment of both transitional and coastal waters. Besides the central functioning role of benthic macrofauna in marine and estuarine ecosystems, various studies often demonstrated its relatively fast response to anthropogenic and natural stresses (Pearson & Rosenberg, 1978; Dauer, 1993). Due to their limited mobility, organisms from benthic communities are quite sensitive to local disturbance, and due to their permanence over seasonal time scales, they integrate the recent history of disturbances, which might not be detected in the water column (Bettencourt et al., 2004). In benthic communities, we can also find different species exhibiting different tolerance levels to stress (Dauer, 1993), which covers the WFD's requirement postulating the integration of sensitive species.

5.1.1.2.1   Basic parameters   To assess the ecological quality of this biological element, WFD requires both abundance and composition to be taken into

consideration (WFD, 2000/60/CE). In Portugal, following the TICOR Project guidelines (Bettencourt *et al.*, 2004), the following metrics were proposed:

A. Abundance. The metrics selected to assess this structural parameter were the Shannon–Wiener (H') and Margalef (d) indices (both described in Section 2.3.3 of Chapter 2), since these indices provide complementary diversity measures. The Shannon–Wiener Index takes proportional abundance of species into account, whereas the Margalef Index focuses on species' richness.

B. Composition. The metric selected to evaluate this parameter was *AMBI* (described in Section 2.3.1 of Chapter 2), which is based on the presence of sensitive species and pollution indicator species.

5.1.1.2.2 Macroinvertebrates assessment method: P-BAT   The chosen indicators provide different outputs, and since WFD requires the ecological status to be reported as an Ecological Quality Ratio (EQR) ranging between 0 and 1, a method combining results from different classifications into a single value had to be adopted. This EQR scale will correspond to five final classes of EQS: Bad, Poor, Moderate, Good and High. This way, the Portuguese Benthic Assessment Tool (P-BAT) was developed to fulfil all WFD requirements (see description in Chapter 2, Section 2.3.5) and assess EQS based on the macroinvertebrate's biological element.

5.1.1.2.3 Data set   In Chapter 4, we have already seen that two of the indices included in this multimetric tool, the Margalef Index and the *AMBI*, could discriminate between areas of the Mondego estuary, namely between both arms of the estuary and between areas with finer sediments and higher organic content, areas affected by dredging and less disturbed areas. However, based on the data set used in Chapter 4, corresponding to 1990, 1992, 1998 and 2000, although a slight improvement of the systems' ecological condition could be detected in 2000, shortly after the application of preliminary mitigation measures, none of the tested indices could detect significant differences between years.

In this chapter, to evaluate the performance of the multimetric tool in tracking impacts and ecological improvements in the Mondego estuary we included data from more recent years (2004 and 2006), which were compared to years of higher disturbance in the system (1990 and 1992), prior to the application of mitigation measures. The three indices (d, H' and *AMBI*) capture different aspects of the benthic communities' structure and therefore, when included in a multimetric approach, they should reflect the state of the system regarding this biological component. At the same time, they should cover for the deficiencies of each other, thereby increasing the overall robustness of the assessment.

For this exercise, 10 subtidal sampling stations located at the lower part of the estuary were chosen (Figure 22), which allow tracking the events previously described in Section 5.1.1. This data set represents two distinct periods regarding disturbance intensity at both arms of the estuary: the early 90s (1990/92) and a *post*-mitigation measures period (2004/6).

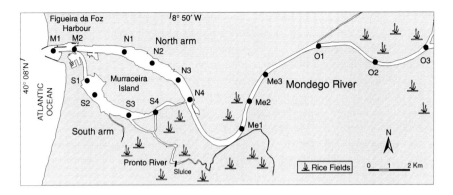

Figure 22. The figure shows the location of the 16 subtidal sampling stations at the Mondego estuary: M1, M2 (mouth); N1, N2 (downstream north arm); N3, N4 (upstream north arm); S1, S2 (downstream south arm); S3, S4 (upstream south arm); Me1, Me2, Me3 (mesohaline zone) and O1, O2, O3 (oligohaline/tidal fresh zone).

In the later period (2004/6), six additional sampling stations located upstream in the estuary were also monitored, which allowed covering the entire estuarine range, from its marine to its freshwater limits (Figure 22).

Stations were sampled during spring season, using van Veen grabs. Five replicates (0.0496–0.078 m², depending on the area of the grab utilised) per sample were collected at subtidal. Samples were sieved through a 1-mm sized mesh and preserved in a 4% buffered formalin solution. Individuals were sorted, counted and identified at species level whenever possible. Taxa in the benthic data matrix are all at species or genus level except for the Chironomidae, Nemertea and Oligochaeta individuals, following truncation rules proposed by the Northeast Atlantic Geographic Intercalibration Group (NEA GIG) (Borja et al., 2007). All macrobenthic abundance data were transformed to number of individuals per square metre (ind m⁻²).

For benthic assessment, the estuary was divided into stretches (Teixeira et al., 2008a), and for each of them, specific benthic reference conditions were established (Table 31).

The reference conditions for the lower estuary (mouth, southern and north arms) followed proposals made by Teixeira et al. (2008b) from a study on the variation of the indices along sixteen years, from 1990 to 2006, at springtime, under different disturbance situations. The reference conditions for the upper estuarine areas (mesohaline and oligohaline/tidal fresh) were settled after studying the indices variation for four consecutive years, from 2003 to 2006, in all seasons. The reference values for each area were settled considering the best value found for each index at each area and increasing it by 20%.

5.1.1.2.4   EQS Results   Table 32 summarises the values of the indices calculated at the 16 sampling stations, during the winter season, in the two periods

Table 31. High and Bad reference values, to evaluate the ecological quality status of benthic macroinvertebrates of distinct areas along the estuarine gradient at the Mondego estuary.

		High reference values						Bad refer-ence values
	Mouth	North arm		South arm		Meso-haline	Oligo-haline/ tidal fresh	
		Down-stream sandy	Upstream sandy	Down-stream sandy	Upstream muddy sand			
d	5	4	3	4	3	3	2	0
H' (Log 2)	4	4	4	4	4	3	2.6	0
AMBI	0.8	1	1	1.5	2.4	2.4	2.4	7

studied (period I: 1990, 1992; period II: 2004, 2006). Lower values for each index were found in period I in the north arm and in period II in the oligohaline zone (d: 0.0; H': 0.0; AMBI: 4.5). The best results for each index were obtained in period II, at the mouth of the estuary (d: 3.1; H': 3.2; AMBI: 1.4).

In all studied years, a general decreasing trend in biodiversity could be observed from the mouth towards the inner areas of the estuary, implying a decrease in the values of d and H' and an increase in AMBI values.

To detect significant ecological differences regarding the ecological condition of benthic assemblages through the temporal disturbance–recovery gradient and the estuarine spatial gradient, a Permutational Multivariate Analysis of Variance (PERMANOVA) (Anderson, 2001; McArdle & Anderson, 2001) was applied to the indices' values. To get a balanced analysis we considered only those stations for which data were available for the entire study period (N1, N2, N3, N4, S1, S2, S3 and S4). Using PERMANOVA software (Anderson, 2005), analyses were performed considering four zones (downstream northern arm, upstream northern arm, downstream southern arm and upstream southern arm) and two periods of distinct pressures (considering the random factor year nested within the period: two levels), with n = two stations per zone × period × year.

Figure 23 shows the variation of the indices for the data considered in this analysis. Similar to what we illustrated in Chapter 4, the present data set reinforces that these three indices show significant differences among distinct areas of the Mondego estuary. They appear to be sensitive to the estuarine gradient and therefore capable of differentiating the inner zones at each arm of the estuary from the more marine-influenced areas. The Margalef Index distinguished the downstream zone of the southern arm from the upstream areas of both arms. Also, in period I, the Shannon–Wiener Index could differentiate between the downstream and upstream zones of the southern arm, and in period II between the downstream and upstream zones at the northern arm. AMBI also differentiated between downstream and upstream areas of the northern arm. In addition, AMBI could differentiate the

Table 32. Ecological indices: winter results at all sampling stations along the Mondego estuary.

Station	d				H'				AMBI			
	1990	1992	2004	2006	1990	1992	2004	2006	1990	1992	2004	2006
M1	–	–	–	1.8	–	–	–	1.6	–	–	–	1.5
M2	1.0	1.2	2.8	3.1	2.1	2.1	3.2	2.0	2.6	3.0	1.4	2.4
N1	0.4	0.4	1.6	1.3	0.8	0.3	2.4	2.6	1.5	3.0	1.9	1.7
N2	0.0	0.7	1.1	1.2	0.0	1.5	1.8	2.1	3.0	2.4	1.3	2.6
N3	0.5	1.2	1.5	0.9	1.2	2.0	1.8	1.3	3.0	2.3	3.0	3.0
N4	0.4	0.9	0.8	1.4	0.9	1.5	0.9	2.2	3.0	2.6	3.0	3.0
S1	1.3	1.3	1.1	1.5	2.3	2.3	1.5	2.5	2.7	1.6	0.7	2.5
S2	1.2	1.5	2.1	2.0	2.2	2.4	2.3	2.5	3.3	3.0	3.0	2.6
S3	0.8	0.8	1.3	1.2	0.8	1.0	1.7	1.9	3.0	3.0	3.0	3.0
S4	1.0	1.8	0.8	1.8	1.6	2.2	1.7	2.6	3.1	3.2	3.0	3.2
Me1	–	–	0.6	0.7	–	–	0.5	0.9	–	–	3.0	3.0
Me2	–	–	0.8	0.5	–	–	1.2	0.6	–	–	4.1	3.1
Me3	–	–	0.2	0.5	–	–	0.9	0.5	–	–	4.0	3.1
O1	–	–	0.0	0.7	–	–	0.0	1.0	–	–	4.5	4.1
O2	–	–	0.0	0.7	–	–	0.0	1.1	–	–	4.5	4.1
O3	–	–	0.2	0.9	–	–	0.4	1.7	–	–	4.4	4.2

Note: Margalef Index (d); Shannon–Wiener Index (H'); AZTI Marine Biotic Index (AMBI).

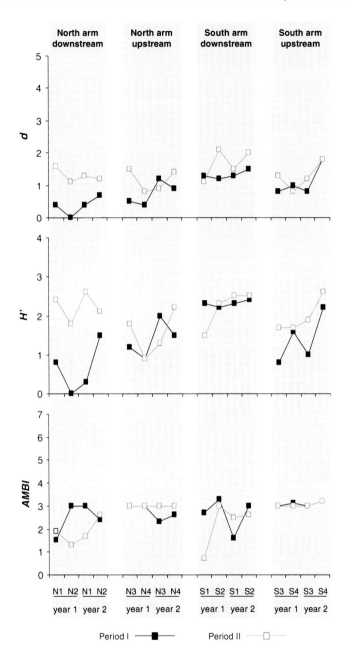

Figure 23. The figure shows the variation of the ecological indices: Margalef (*d*),
Shannon–Wiener (*H'* bits indiv^{-1}) and *AMBI* in four estuarine zones in
period I (year 1: 1990; year 2: 1992), under disturbance; and in period
II (year 1: 2004; year 2: 2006), after experimental mitigation measures
at the estuary in 1997/98.

upstream zone at the southern arm, characterised by sediments with finer grain size and higher percentage of organic matter (Teixeira *et al.*, 2008b), from the other three estuarine zones during the whole study period.

Nevertheless, the comparison of the indices' values obtained for the northern and southern arms, before and after the application of experimental mitigation measures, showed that the three indices did not have the same capability to capture the changes occurred in the system. The Margalef Index detected significant differences between the two periods in the entire estuary and in all four zones. The Shannon–Wiener Index demonstrated a significant interaction between *zone versus period*, but the post-hoc tests could not identify those differences. A posterior one-way ANOVA, regarding periods before and after the application of mitigation measures and considering each zone separately, revealed that this index only detected significant differences between period I and II at the downstream stations of northern arm. Finally, *AMBI* did not detect any significant differences between the two studied periods (Figure 23).

But what is the picture regarding the benthic ecological condition through time if we consider now the final ecological classification – EQS – for which the indices are brought together into a multimetric tool?

The EQS of the Mondego estuary benthic communities, given by P-BAT tool, indicated an overall variation from Poor/Moderate in the early 1990s to Moderate/Good in the period after the application of mitigation measures (Table 33).

The multimetric tool therefore seems more sensitive, than each index by its own, to the disturbance–recovery gradient observed in each arm of the estuary in the two periods.

In period I, stations at the mouth were in a Moderate status. In this same period, the downstream area of the northern arm, which was intensively physically disturbed in the first half of the 1990s, was classified as Bad/Poor at that time, while its upstream areas were classified as Poor/Moderate in 1990/92 and became more affected in 1993/94. In period II, the area of the mouth presented Good status and the entire northern arm was classified as Moderate, which indicates an improvement in ecological conditions.

The classifications obtained for the northern arm of the system (stations M1 to N4) were coherent with our understanding of the system, although some dredging was still occurring in period II, especially at the downstream area. The induced disturbance level was not comparable to the physical interventions that took place in period I. In both periods, the mouth area always presented a classification one class above that of the rest of the stations along the main navigation channel. The downstream limit of the estuary is in fact a very dynamic zone, where the proximity of coastal environments may contribute to the benthic communities' higher recovery rate after disturbance, due to colonisation by marine species.

At the southern arm, a gradual improvement of the ecological status of the benthic communities could be observed, from Moderate in 1990 to Good in 2006. In the beginning of 2004, some Moderate classifications were still recorded among some southern arm stations, which might have to do with the slow recovery of the benthic communities (Borja *et al.*, 2003b).

Table 33. P-BAT ecological classification (EQR and EQS) of the benthic macroinvertebrates of the Mondego estuary in the two periods studied (1990/92 and 2004/6) at each estuarine area.

Water bodies (Ferreira et al., 2006)	Estuarine area	Station	Heavily disturbed period				Slightly disturbed period			
			1990		1992		2004		2006	
			EQR	EQS	EQR	EQS	EQR	EQS	EQR	EQS
WB1	Mouth	M1	–	–	–	–	–	–	0.61	G
		M2	0.52	M	0.50	M	0.79	G	0.64	G
WB2	Downstream north arm	N1	0.34	P	0.24	P	0.58	M	0.57	M
		N2	0.17	B	0.39	P	0.51	M	0.49	M
	Upstream north arm	N3	0.41	M	0.59	M	0.55	M	0.46	M
		N4	0.38	P	0.50	M	0.41	M	0.58	M
WB4	Downstream south arm	S1	0.55	M	0.61	G	0.57	M	0.60	G
		S2	0.50	M	0.56	M	0.60	G	0.64	G
	Upstream south arm	S3	0.50	M	0.52	M	0.62	G	0.62	G
		S4	0.57	M	0.68	G	0.57	M	0.71	G
Not defined for these stretches	Mesohaline	Me1	–	–	–	–	0.46	M	0.51	M
		Me2	–	–	–	–	0.45	M	0.45	M
		Me3	–	–	–	–	0.36	P	0.44	M
	Oligohaline/Tidal fresh	O1	–	–	–	–	0.20	P	0.46	M
		O2	–	–	–	–	0.29	P	0.57	M
		O3	–	–	–	–	0.20	P	0.41	M

Note: Ecological quality status (EQS): H, High; G, Good; M, Moderate; P, Poor and B, Bad. '–', no data available.

Despite the strong eutrophication symptoms that occurred in the southern arm in the early/mid 1990s, intertidal benthic communities were far more affected than the subtidal ones (Dolbeth *et al.*, 2007; Patrício *et al.*, submitted). On the contrary, the physical disturbance in the north arm directly affected the subtidal habitats, justifying therefore the lowest ecological classifications obtained in this arm with regard to the benthic communities.

Monitoring of the mesohaline to the oligohaline/tidal fresh ranges in this estuary took place only from 2003 onwards, and therefore little knowledge has already gathered regarding the benthic communities in these areas. Nevertheless, the natural impoverishment that characterises these stretches seems to be reflected in the low classifications obtained using the P-BAT tool (Table 33). The prolongation of monitoring in these areas will clarify whether the proposed reference conditions are adequate.

In general, the EQS classifications produced by this multimetric tool are satisfactory and correspond to our information of the system regarding the evolution of the subtidal benthic communities through time (Teixeira *et al.*, submitted).

An issue that is worth discussing here is with regard to the time of the year chosen to evaluate the ecological quality of benthic macroinvertebrate communities. The choice of the winter season had to do with two main reasons. First, since reference conditions should cover the spectrum of natural variability (WFD, 2000/60/CE), the winter season was chosen to set reference values for high quality standards. In winter, the biological activity is minimum (Marques, 2001), so the indices could reflect the lowest natural condition within an eventually Good ecological state, which is the ideal situation to set the boundaries for each of the five ecological status classes. Second, monitoring this biological element in winter would avoid recruitment events that occur in spring and summer seasons, which might cause many indices to react to this natural variation in the same way as to an anthropogenic disturbance (Beukema, 1988; Dauer *et al.*, 1993; Teixeira *et al.*, 2007).

The synergistic effects of recruitment, food availability, water temperature, predation and stress induced by hydrodynamics, among others, are beneath the seasonal variability observed in the benthic communities (Reiss & Kröncke, 2005). These events will regulate benthic abundance, diversity and community structure, parameters to which the univariate indices, such as the *d* and *H'* included in the P-BAT tool, are more sensitive (Reiss & Kröncke, 2005). Indices based on general life history traits, such as the *AMBI* included in the P-BAT tool, seem to be less influenced by the seasonal variability of the macrofauna (Reiss & Kröncke, 2005). Thus, the effect of seasonal variability must be accounted for during the process of EQS assessment, and until further research on this matter, the best season to track the impacts on benthic communities is controversial.

Another issue to be discussed is the adjustment of the boundaries between the five classes of EQS which, for the present exercise, have been assumed as equidistant along the 0 to 1 EQR. The classifications obtained must be further checked for agreement with the WFD's normative definitions, and subject to the intercalibration (IC) exercise between metrics of different member states, which has not yet been completed for transitional waters.

5.1.1.3 Integration of different biological quality elements' classification into a final EQS for the Mondego estuary

To proceed with the ecological assessment of this system in accordance with the WFD, the different biological quality elements must be integrated. In our example, we deal only with benthos and macroalgae, utilising the data available provided by monitoring in recent years. The resulting classifications per biological element and per water body are shown in Figure 24.

In our opinion, one of the possible strengths of taking into consideration several biological elements is to account for the distinct response of each ecosystem component following a disturbance event (namely, nutrient enrichment). It is well known that phytoplankton may show a faster reaction than benthic macroinvertebrates and, similarly, opportunistic macroalgae react faster than rooted macrophytes.

For example, in this first case study, during the 1990s, there was an increase in water turbidity, a greater availability of nutrients for consumption by primary

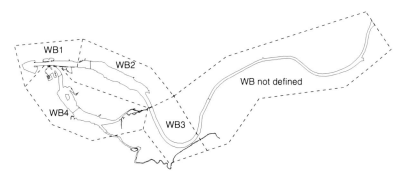

Water Body	Phytoplankton	Other aquatic flora[a]	Benthic invertebrate fauna	Fish fauna
WB1	--	n.a.	**Good**	--
			2006	
WB2	--	--	**Moderate**	--
			2006	
WB3	--	n.a.	**Moderate**	--
			2006	
WB4	--	**High**	**Good**	--
		2000	2006	

(a) denotes classification based only on the opportunistic macroalgae macroalgal assessment; . n.a., not applicable; '–' not monitored.

Figure 24. The figure shows water bodies at the Mondego estuary (WB1, WB2, WB3, WB4) as defined by Ferreira *et al.* (2006), and the corresponding monitoring date and classifications of the biological quality elements.

producers and an increase in water residence time in the southern arm (due to the upstream reduced inflow), which favoured immediately the occurrence of opportunistic green macroalgal blooms. Nevertheless, only later, the *Z. noltii* meadows started to decline, most probably as a function of competition with macroalgae (Marques *et al.*, 2003). Therefore, evaluating separately these two biological elements, opportunistic macroalgae may be used as 'early warning' signs regarding nutrient's pressure.

According to WFD, an integration of all biological assessments with information on physicochemical and hydromorphological conditions must be undertaken to produce a final ecological status classification for each water body (Figure 25). Moreover, in accordance with WFD guidelines, the final ecological status of a water body should be determined by the lower status of either the biological quality elements or the physicochemical elements (Vincent *et al.*, 2002). However, this *'one out, all out principle'* has been widely questioned. Due to different sampling frequencies, the high spatial and temporal variability of some biological elements and the role of some of the elements as good indicators, any form of considering the data should be discussed, to derive a more accurate global classification (Borja *et al.*, 2004). In the Mondego estuary, for example, a much more extensive monitoring programme has been carried out for benthos (since 1985) or macroalgae (since 1993), than for fishes (since 2001) or for phytoplankton (since 2004). Therefore, the degree of confidence when interpreting monitoring results in the scope of the WFD is not balanced between all biological elements assessed in this system.

5.1.2   Coastal waters: The Buarcos Bay case study

The Buarcos Bay (40°10′5.99″N, 8°53′22.27″W) (Figure 26) belongs to type A5 from theCW Portuguese typology (Bettencourt *et al.*, 2004), which was included in the NEA1 typology (van de Bund *et al.*, 2004) within the North-East Atlantic Ocean area (WFD, 2000/60/CE). These are open and exposed CW, euhaline, mesotidal with amplitudes varying from 1 to 3 m, and frequently having natural turbidity and nutrient enrichment due to the occurrence of upwelling phenomena (Ambar & Dias, 2008). In the summer, the main currents affecting the Portuguese coast are the Canary Current, which has a strong southward flow (12 cm s^{-1}) originating from the north, and the Azores Current, which enters the region in the south, establishing a west to east circulation. In winter, the Azores Current has twice the velocity and there is little circulation of seawater in the region. The circulation of seawater along the Iberian Coast is predominantly south to north, with velocities of 1.6 cm s^{-1}.

The Buarcos Bay as mentioned above for the city of Figueira da Foz, is strongly influenced by summer tourism. However, it is also affected by other anthropogenic activities existing inside the Mondego estuary because it is located northward adjacent to its mouth. This rocky shore comprises extensive platforms, with irregular surface and abundant basic pools (Figure 27). Biota is characterised by a green belt at higher zones and more abundant reds and browns in lower areas. Macrofauna is dominated by mussels, barnacles, limpets and periwinkles.

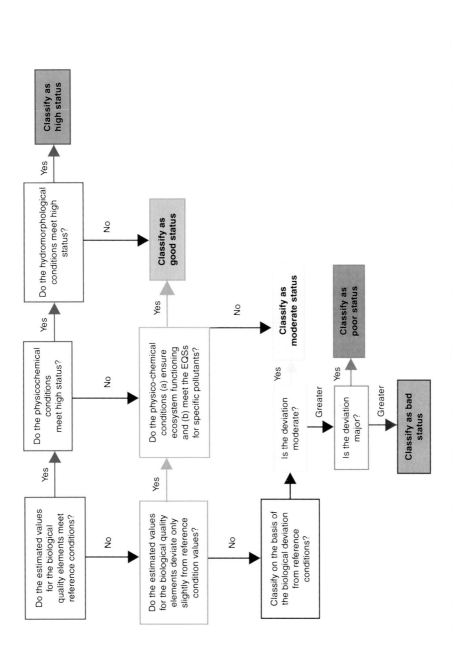

Figure 25. The figure gives an indication of the relative roles of biological, hydromorphological and physicochemical quality elements in ecological status classification according to the normative definitions in WFD Annex V 1.2 (from Vincent *et al.*, 2002).

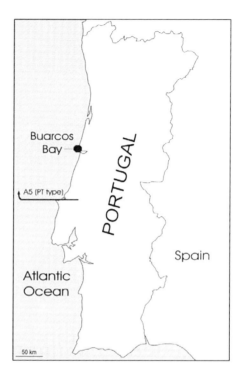

Figure 26. The figure shows the Buarcos Bay sampling area.

Figure 27. The figure shows the Buarcos Bay rocky shore.

5.1.2.1   Marine macroalgae EQS assessment

5.1.2.1.1   Basic parameters   The WFD indicates 'Other Aquatic Flora', where the macroalgae are included, as a key biological element to assess the ecological status of CW. Intertidal macroalgal communities can be characterised according to several factors such as the species composition, diversity among green, red and brown classes, and the biomass of some present *taxa*. These attributes, analysed individually or as a combination, can be used to monitor the proper functioning of aquatic systems and infer about their ecological status (Schramm, 1999; Orfanidis *et al.*, 2003; Krause-Jensen *et al.*, 2007; Scanlan *et al.*, 2007; Wells *et al.*, 2007; Juanes *et al.*, 2008; Neto *et al.*, submitted). The measurable attributes selected here constitute the Portuguese Marine Macroalgae Assessment Tool (P-MarMAT), and are the following: macroalgal species richness, proportion of Chlorophyceae (greens), number of Rhodophyceae (Reds), Ecological Status Groups (ESG) ratio (Orfanidis *et al.*, 2001, 2003, Wells *et al.*, 2007), proportion of opportunists, coverage of opportunists and shore description. It is also necessary to elaborate the selected species list (SSL), representative of local macroalgal natural occurrences, according to Neto *et al.* (submitted).

5.1.2.1.2   Marine macroalgae assessment method: P-MarMAT   Based on the aforementioned basic parameters, the P-MarMAT was developed to fulfil all the WFD requirements (see description in Chapter 2, Section 2.3.5) and set the EQS based in the marine macroalgae biological element. This tool was applied to the following data set.

5.1.2.1.3   Data set   In early autumn of 2007, the Buarcos rocky shore was sampled. Transects perpendicular to the water line were defined over the intertidal rocky substratum, and in a regular grid (every metre) along transects, a perpendicular set of three replicate digital photographs was made for later analysis of macroalgal coverage. The total number of photographic records depended on the shore length. Simultaneously, all macroalgal *taxa* observed at shore were registered on occurrence field tables, and a detailed description of shores was also made for later use in the assessment method as a correction factor over the species richness (see Table 34).

5.1.2.1.4   EQS results   Results from the application of P-MarMAT in assessing the ecological quality of the Buarcos shore are shown in Table 35. It is possible to see how to come from the first steps of achieved 'metrics values', through the attributed 'scores', to the final definition of 'EQR' and respective 'EQS'. The Buarcos Bay shore, agreeing with our expectations, was classified as Good.

P-MarMAT is a methodology compliant with WFD, considering 'abundance' and 'composition' of macroalgae. Although strongly based on scientific knowledge, the methodology is easy to use (by non-experts with reasonably high taxonomical experience), with a quick outputting of assessment results and considerably low cost. The majority of the metrics originated in the RSL concept (Wells *et al.*, 2007), since this is also based on a list of species, easy to identify by

Table 34. Field sampling sheet to record basic shore descriptions (adapted from Wells *et al.*, 2007).

General information				
Shore name		Date		
Water body		Tidal height		
Grid ref.		Time of low tide		
Shore descriptions				
Presence of turbidity	Yes = 0	Sand scour	Yes = 0	No = 2
(known to be non-anthropogenic)	No = 2	Chalk shore	Yes = 0	No = 2
Dominant shore type		Subhabitats		
Rock ridges/ outcrops/platforms	= 4	Wide shallow rock pools		
Irregular rock	= 3	(> 3 m wide and < 50 cm deep)	= 4	
Boulders large, medium and small	= 3	Large rockpools (> 6 m long)	= 4	
Steep/vertical rock	= 2	Deep rockpools (50% > 100 cm deep)	= 4	
Non-specific hard substrate	= 2	Basic rockpools	= 3	
Pebbles/stones/small rocks	= 1	Large crevices	= 3	
Shingle/gravel	= 0	Large overhangs and vertical rock	= 2	
		Others habitats (please specify)	= 2	
Dominant biota		Caves		
Ascophyllum		None		
Fucoid				
Rhodophyta mosaics		Total number of subhabitats		
Chlorophyta		> 4 3 2 1 0		
Mussels				
Barnacles		General comments		
Limpets				
Periwinkles				

Table 35. P-MarMAT assessment results for Buarcos Bay in autumn 2007.

		Buarcos
Metrics values	Species richness	26
	Proportion of greens	12
	Number of reds	18
	ESG ratio	2.3
	Proportion of opportunists	9
	Coverage of opportunists	75
	Shore description	14
Scores	Species richness	4
	Proportion of greens	3
	Number of reds	3
	ESG ratio	3
	Proportion of opportunists	3
	Coverage of opportunists	0
	Shore description	2
	Final score	22
	EQR	0.61
	EQS	Good

Note: Values for each of the metrics, scores, ecological quality ratio (EQR) and ecological quality status (EQS).

non-expert people with good algal-identification training, and in a pool of other metrics calculated from the *taxa* present there. Additionally, the methodology was complemented by including a coverage value to fulfil the abundance parameter stated in the WFD document. The SSL, includes the more representative and easy-to-identify macroalgal taxa. It is in agreement with classes' proportions (Chlorophyceae, Phaeophyceae and Rhodophyceae) present at intertidal rocky seashores, and should be adapted to macroalgal taxa present at shores supposed to be assessed.

Buarcos Bay was assessed for this example in early autumn, although our experience suggests, in agreement with what is stated in WFD, that it should be done between late spring and early summer. This is the period of maximum development for most seaweed populations in temperate seas, and it avoids the seasonal episodic explosion of ephemeral species occurring in April–May and the worst weather conditions that can usually affect sampling activities in littoral areas during autumn and winter.

Although the boundaries adopted here were equidistant (0.20), it corresponds to the initial proposal of the methodology and should be improved in future. A comparison and IC between P-MarMAT and other existent methodologies (e.g. RSL, CFR) should be done to fine-tune quality classes limits and be sure about their compliance in assessment concerns. An earlier comparison of RSL and CFR methodologies was done under the umbrella of the IC exercise (Guinda *et al.*, 2008), and it was concluded that the latter could be more accurate in assessing the quality of Iberian shores. The reason is the inclusion of the coverage of opportunists metric, which allows following discrepancies existing on normal distribution patterns of macroalgae. These reasons together with our findings and results increase optimism on a future good IC exercise, even more when a previous draft version of P-MarMAT (including 'proportion of reds' instead of the present 'number of reds') was intercalibrated against CFR, producing an agreement value of 0.89 ('excellent') from Kappa analysis (WFD IC technical report 2007, Part 3 – Coastal and Transitional Waters, Section 4 – Macroalgae. http://circa.europa.eu/Public/irc/jrc/jrc_eewai/library?l=/milestone_reports/ milestone_reports_2007/coastaltransitional/coast_nea_gig/macroalgae_final/ _EN_1.0_&a=d).

5.1.2.2   Benthic macroinvertebrates EQS assessment
5.1.2.2.1   Basic parameters   The same methodology as described in Sections 5.1.1.2.1.
5.1.2.2.2   Benthic macroinvertebrates assessment method: P-BAT   The same methodology as described in Sections 5.1.1.2.2.

5.1.2.2.3   Data set   Buarcos Bay was surveyed twice in different seasons: winter 2006 and summer 2006. The subtidal soft bottom community samples were collected with a Smith–McIntyre grab (0.1 m^2 area) along a transect perpendicular to coast line at 10 (summer), 20 (winter and summer) and 30 (winter) meters depth. Three replicates per sample were taken, except in the 30-m survey where only two replicates were available.

Biological material was preserved in neutralised 4% formalin solution prepared with marine water. Samples were sieved through a 1-mm mesh and stored in alcohol (70% ethanol) after sorting. Individuals were counted and identified at species level whenever possible. Taxa in the benthic data matrix are all at species or genus level except for the Chironomidae, Nemertea and Oligochaeta individuals, following truncation rules proposed by the NEA GIG (Borja *et al.*, 2007).

All macrobenthic abundance data were transformed to number of individuals per square metre (ind m^{-2}).

Reference conditions at the NEA1 type in Portugal, for the metrics of P-BAT methodology (Table 36), were settled after surveys in the scope of RECITAL project along the Portuguese coast. The methodology for CW has already been the subject of IC between member states within the NEA GIG, and therefore the thresholds between the five EQS classes (Table 36) are already accepted for this type (WFD – NEA GIG report section 2 (benthic inverts) Final draft June 07 http://circa.europa.eu/Public/irc/jrc/jrc_eewai/library?l=/milestone_reports/ milestone_reports_2007/coastaltransitional/coast_nea_gig&vm=detailed&sb= Title).

Table 36. Reference conditions at NEA1 type of Portuguese coastal waters, for indices (Margalef, Shannon–Wiener and *AMBI*) in the P-BAT multimetric.

P-BAT: Portuguese benthic assessment tool CW NEA1			EQS and EQR boundaries	
Margalef	Shannon–Wiener	*AMBI*		
5	**4**	**0**	High	1
			/	0.79
			Good	
			/	0.58
			Moderate	
			/	0.44
			Poor	
			/	0.27
0	**0**	**7**	Bad	0

Note: Reference values for High and Bad ecological status (EQS) are presented along with EQR boundaries between the five EQS classes, after the IC setting boundaries procedure.

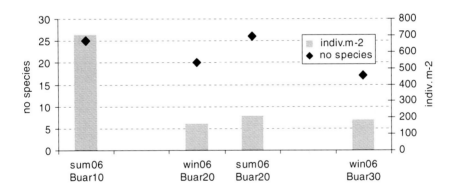

Figure 28. The figure shows the variation of the number of species and density along the depth gradient (10–30 m) in Buarcos Bay, in the 2006 surveys (winter and summer).

5.1.2.2.4 EQS results Some relevant considerations about the methodology were already made in Section 5.1.1. (Mondego estuary), which are still valid for this section of CW. After this remark, we may show results from Buarcos Bay assessment. Figure 28 shows the variation of community structural parameters such as the number of species and the density of individuals along a depth gradient in Buarcos Bay.

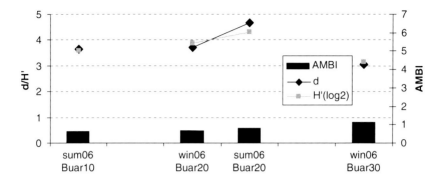

Figure 29. The figure shows the variation of the three indices of the P-BAT multimetric along the depth gradient (10–30 m) in Buarcos Bay, in the 2006 surveys (winter and summer).

Table 37. EQR and EQS results along the depth gradient (10–30 m) in Buarcos Bay, in the 2006 surveys (winter and summer).

Depth	Year 2006	EQR	EQS
10 m	Summer	0.85	H
20 m	Winter	0.87	H
	Summer	0.97	H
30 m	Winter	0.75	G

The number of individuals was markedly higher at 10 m depth than for 20 or 30 m, and the number of species was more dependent on season, increasing in summer, as expected.

Figure 29 shows the variation of the three indices included in the P-BAT multimetric tool. It can be seen that Margalef and Shannon–Wiener indices reached higher values during summer, but no clear tendency could be observed from depth variation. Higher values from *AMBI*, meaning lower quality, were obtained at 30 m depth.

By analysing the P-BAT assessment results, one can say that the ecological indices in Buarcos Bay reached higher values, near the reference conditions initially proposed for the NEA1 typology, reflecting a community with very minor disturbance. The final EQR obtained for this area varied between 0.75 and 0.97, indicating a community between Good and High EQS (Table 37).

5.1.2.3   Integration of different biological quality elements' classification into a final EQS for the Buarcos Bay

As stated previously in Section 5.1.1.3., to proceed with the ecological assessment of this system in accordance with the WFD, the different biological quality

Table 38. Buarcos Bay type as defined by Ferreira *et al.* (2006).

Type	Phytoplankton	Other aquatic flora[a]	Benthic invertebrate fauna
A5/NEA1	–	**Good**	**High/Good**
		2007	2006

Note: EQS based on each biological quality element. n.a., not applicable. '–', not monitored.
[a]Based only on the marine macroalgae assessment.

elements must be integrated. Also in our CW example, we deal only with benthos and macroalgae. The resulting classifications per biological element is shown in Table 38.

In this example, both elements showed Good quality status, and the problems arising from different classifications coming from different elements is not a concern here. This fact is obviously linked to the generally healthy environment present in the bay, and no tendency of translating a lower quality from any of the considered biological elements is visible. The final classification, bearing in mind these two elements, can be seen as being in Good quality.

5.2   Final remarks

Some conclusions can be derived from the examples presented in the text. Bearing in mind the methodologies explored here, one may agree that multimetric methodologies are able to assess the ecological quality of transitional and coastal waters to a considerable extent. Although some improvements are already expected to occur in the following years, scientists are committed to contribute to the development of assessing tools compliant with WFD demands. This is applicable to the methodologies presented here to assess marine macroalgae, the P-MarMAT, and the benthic macroinvertebrates, the P-BAT. We feel that both have the potential to be applied in other geographic áreas, but like others, some adaptations should be done, namely on the SSL, reference conditions or boundaries.

In general, the selected metrics reacted as expected along the disturbance gradient, revealing worst quality conditions where stronger impacts were present. These works reinforce the idea that macroalgae and macroinvertebrates can integrate the effects of environmental conditions. These are sessile organisms or present restricted mobility; hence they serve as good biological sensors to evaluate water quality, which is not always the case of evaluations based on instant physicochemical parameters.

Acknowledgements

The present work was prepared in the scope of the research projects WADI (INCO-CT-2005-015226) and WISER (FP7-ENV-2008-1), funded by EU, RECONNECT (PTDC/MAR/64627/2006), funded by FCT (Portuguese National Board Scientific Research) and RECITAL (contract no. 2005/056/INAG) funded by INAG – National Water Institute (Portugal). Additionally it benefited from three grants (SFRH/BPD/26604/2006, SFRH/BPD/41701/2007, SFRH/BD/24430/2005) attributed by FCT. The authors are indebted to C. Marcos and A. Pérez-Ruzafa, colleagues, but above all friends, from the University of Múrcia, who made available the extensive sets of data on the Mar Menor lagoon, Escombreras basin and Cape Tiñoso utilized in Chapter 4.

References

Abou-Aisha, K.H., Kobbia, I.A., Abyad, M.S., Shabana, E.F. & Schanz, F. 1995. Impact of phosphorous loadings on macro-algal communities in the Red sea coast of Egypt. *Water, Air and Soil Pollution*, 83: 285–297.

AFNOR. 1992. Norme française NF T 90–350. Détermination de l'Indice Biologique Global Normalisé (IBGN), 9 pp.

AFNOR. 2000. Norme française NF T 90–354. Qualité de l'eau. Détermination de l'Indice Biologique Diatomées (IBD), 63 pp.

Agard, J.B., Gobin, J. & Warwick, R.M. 1993. Analysis of marine macrobenthic community structure in relation to pollution, natural oil seepage and seasonal disturbance in a tropical environment (Trinidad, West Indies). *Marine Ecology Progress Series*, 92: 233–243.

Alba-Tercedor, J. 1996. *Macroinvertebrados acuáticos y calidad de las aguas de los rios. IV Simposio del Agua en Andalucia (SIAGA)*, Almeria, España, pp. 203–213.

Alba-Tercedor, J., Jaimez-Cuéllar, P., Álvares, M., Avilés, J., Bonada, N., Casa, J., Mellado, A., Ortega, M., Pardo, I., Prat, N. & Rieradevall, M. 2002. Caracterización del estado ecológico de rios mediterrâneos ibéricos mediante el índice IBMWP (antes BMWP'). *Limnetica*, 21(3–4): 175–185.

Alba-Tercedor, J. & Pujante, A. 2000. Running-water biomonitoring in Spain. Opportunities for a predictive approach. In: Wright, J.F., Sutcliffe, D.W. & Furse, M. (Eds.), *Assessing the Biological Quality of Freshwater: RIVPACS and Similar Techniques*. Freshwater Biological Association-Special Publications, Cumbria, UK, pp. 207–216.

Alba-Tercedor & Sánchez, O.A. 1988. Un método rápido y simples para evaluar la calidad biológica de las águas correntes basado en el de Hellawell (1978). *Limnética*, 4: 51–56.

Albrecht, W.N., Woodhouse, C.A. & Miller, J.N. 1981. Nearshore dredge – spoil dumping and cadmium, copper and zinc levels in a dermestid shrimp. *Bulletin of Environmental Contamination and Toxicology*, 26: 219–223.

Ambar, I. & Dias, J. 2008. Remote sensing of coastal upwelling in the North-Eastern Atlantic Ocean. In: Barale, V. & Gade, M. (Eds.), *Remote Sensing of the European Seas*. Springer, Netherlands, pp. 141–152.

Anderlini, V.C. & Wear, R.G. 1992. The effects of sewage and natural seasonal disturbance on benthic macrofaunal communites in Fitzray Bay, Wellington, New Zealand. *Marine Pollution Bulletin*, 24: 21–26.

Andersen, A.N. 1997. Using ants as bioindicators: multiscale issues in ant community ecology. *Conservation and Ecology* (online). http://consecol.org/vol1/iss1/atr8.

Anderson, M.J. 2001. A new method for non-parametric multivariate analysis of variance. *Australian Ecology*, 6: 32–46.

Anderson, M.J. 2005. *PERMANOVA: a FORTRAN computer program for permutational multivariate analysis of variance*. Department of Statistics, University of Auckland, New Zealand.

Andrade, F. 1986. *O estuário do Mira: caracterização geral e análise quantitativa da estrutura dos macropovoamentos bentónicos*. PhD Thesis, University of Lisbon, Portugal.

Anger, K. 1975. On the influence of sewage pollution on inshore benthic communities in the south of Kiel Bay. 2. Quantitative studies on community structure. *Helgoländer. Wissenschaftliche Meeresuntersuchungen*, 27: 408–438.

Armitage, P.D., Moss, D., Wright, J.F. & Furse, M.T. 1983. The performance of a new biological water quality score system based on macroinvertebrates over a wide range of unpolluted running water sites. *Water Research*, 17: 333–347.

Baird, D., McGlade, J.M. & Ulanowicz, R.E. 1991. The comparative ecology of six marine ecosystems. *Philosophical Transactions of the Royal Society of London B*, 333: 15–29.

Bald, J., Borja, A., Muxika, I., Franco, J. & Valencia, V. 2005. Assessing reference conditions and physico-chemical status according to the European Water Framework Directive: a case-study from the Basque Country (Northern Spain). *Marine Pollution Bulletin*, 50: 1508–1522.

Bazzaz, F.A. 1986. Life history of colonizing plants: some demographic, genetic, and physiological features. In: Mooney, H.A. & Drake, J.A. (Eds.), *Ecology of Biological Invasions of North America and Hawaii*. Springer-Verlag, Berlin, pp. 96–110.

Bellan, G. 1980. Annélides polychétes des substrats solids de troits mileux pollués sur les côrtes de Provence (France): Cortiou, Golfe de Fos, Vieux Port de Marseille. *Téthys*, 9(3): 260–278.

Bellan-Santini, D. 1980. Relationship between populations of amphipods and pollution. *Marine Pollution Bulletin*, 11: 224–227.

Belsher, T. 1982. Measuring the standing crop of intertidal seaweeds by remote sensing. In: Last, F.T., Hotz, M.C. & Bell, B. (Eds.), *Land and Its Uses Actual and Potential: An Environmental Appraisal*. NATO Seminar on Land and its Uses, Edinburgh 19 Sep–1 Oct 1982, 10, pp. 453–456.

Belsher, T. & Boudouresque, C.F. 1976. L'impact de la pollution sur la fraction algale des peuplements benthiques de Méditerranée. *Atti Tavola Rotonda Internazionale, Livorno*, pp. 215–260.

Bergen, M., Cadien, D., Dalkey, A., Montagne, D.E., Smith, R.W., Stull, J.K., Velarde, R.G. & Weisberg, S.B. 2000. Assessment of benthic infaunal condition on the mainland shelf of Southern California. *Environmental Monitoring and Assessment*, 64: 421–434.

Berger, W.H. & Parker, F.L. 1970. Diversity of planktonic Foraminifera in deep sea sediments. *Science*, 168: 1345–1347.

Best, M.A., Wither, W.A. & Coates, S. 2007. Dissolved oxygen as physico-chemical supporting element in the Water Framework Directive. *Marine Pollution Bulletin*, 55: 53–64.

Bettencourt, A., Bricker, S.B., Ferreira, J.G., Franco, A., Marques, J.C., Melo, J.J., Nobre, A., Ramos, L., Reis, C.S., Salas, F., Silva, M.C., Simas, T. & Wolff, W. 2004. *Typology and reference conditions for Portuguese transitional and coastal waters*. INAG. IMAR. p. 99.

Beukema, J.J. 1988. An evaluation of the ABC method (abundance/biomass comparison) as applied to macrozoobenthic communities living on tidal flats in the Dutch Wadden Sea. *Marine Biology*, 99: 425–433.

Blair, R.B. 1996. Land use and avian species diversity along an urban gradient. *Ecological Applications*, 6: 506–519.

Blanchet, H., Lavesque, N., Ruellet, T., Dauvin, J.C., Sauriau, P.G., Desroy, N., Desclaux, C., Leconte, M., Bachelet, G., Janson, A.-L., Bessineton, C., Duhamel, S., Jourde, J., Mayot, S., Simon, S. & de Montaudouin, X. 2008. Use of biotic indices in semi-enclosed coastal ecosystems and transitional waters habitats – implications for the implementation of the European Water Framework Directive. *Ecological Indicators*, 8(4): 360–372.

Bongers, T. 1999. The Maturity Index, the evolution of nematode life history traits, adaptive radiation and cp-scaling. *Plant and Soil*, 212: 13–22.

Bongers, T., Alkemade, R. & Yeates, G.W. 1991. Interpretation of disturbance-induced maturity decrease in marine nematode assemblages by means of the Maturity Index. *Marine Ecology Progress Series*, 76: 135–142.

Bonne, W., Rekecki, A. & Vincx, M. 2003. Impact assessment of sand extraction on subtidal sandbanks using macrobenthos. In: *Benthic copepod communities in relation to natural and anthopogenic influences in the North Sea*. Ph.D Thesis of W. Bonne, Ghent University, Biology Department, Marine Biology Section, Belgium, pp. 207–226.

Borja, A. 2006. The new European Marine Strategy Directive: difficulties, opportunities, and challenges. *Marine Pollution Bulletin*, 52: 239–242.

Borja, A., Dauer, D.M., Díaz, R., Llansó, R.J., Muxika, I., Rodríguez, J.G. & Schaffner, L. 2008. Assessing estuarine benthic quality conditions in Chesapeake Bay: a comparison of three indices. *Ecological Indicators*, 8(4): 395–403.

Borja, Á., Franco, J. & Pérez, V. 2000. A Marine biotic index to establish the ecological quality of soft-Bottom Benthos within European Estuarine and coastal environments. *Marine Pollution Bulletin*, 40(12): 1100–1114.

Borja, Á., Franco, J. & Muxika, I. 2003c. Classification tools for marine ecological quality assessment: the usefulness of macrobenthic communities in an area affected by a submarine outfall. *ICES CM 2003/Session J-02*, Tallinn (Estonia), 24–28 September.

Borja, A., Franco, J., Valencia, V., Bald, J., Muxika, I., Belzunce, M.J. & Solaun, O. 2004. Implementation of the European water framework directive from the Basque country (northern Spain): a methodological approach. *Marine Pollution Bulletin*, 48: 209–218.

Borja, Á., García de Bikuña, B., Blanco, J.M., Agirre, A., Aierbe, E., Bald, J., Belzunce, M.J., Fraile, H., Franco, J., Gandarias, O., Goikoetxea, I., Leonardo, J.M., Lonbide, L., Moso, M., Muxika, I., Pérez, V., Santoro, F., Solaun, O., Tello, E.M. & Valencia, V. 2003a. *Red de Vigilancia de las masas de agua superficial de la Comunidad Autónoma del País Vasco. Tomo 1: Metodologías utilizadas*. Departamento de Ordenación del Territorio y Medio Ambiente, Gobierno Vasco, 199 p.

Borja, A., Josefson, A.B., Miles, A., Muxika, I., Olsgard, F., Phillips, G., Rodríguez, J.G. & Rygg, B. 2007. An approach to the intercalibration of benthic ecological status assessment in the North Atlantic ecoregion, according to the European Water Framework Directive. *Marine Pollution Bulletin*, 55: 42–52.

Borja, Á., Muxika, I. & Franco, J. 2003b. The application of a Marine Biotic Index to different impact sources affecting soft-bottom benthic communities along European coasts. *Marine Pollution Bulletin*, 46: 835–845.

Bratton, S.P., Hapeman, J.R. & Mast, A.R. 1994. The Lower Susquehanna River Gorge and floodplain (U.S.A.) as a riparian refugium for vernal, forest-floor herbs. *Conservative Biology*, 8: 1069–1077.

Brock, V. 1992. Effects of mercury on the biosynthesis of porphyrins in bivalve molluscs (*Cerastoderma edule* and *C. lamarcki*). *Journal of Experimental Marine Biology and Ecology*, 164: 17–29.

Browder, S.F., Johnson, D.H. & Ball, I.J. 2002. Assemblages of breeding birds as indicators of grassland condition. *Ecological Indicators*, 2: 257–270.

Brown, A.C. 1985. The effects of crude oil pollution on marine organisms: a literature review in the South African context: conclusions and recommendations. *South African National Scientific Progress Report* No. 99, 33 p.

Bryan, G.W. & Gibbs, P.E. 1987. Polychaetes as indicators of heavy-metal availability in marine deposits. In: Capuzzo, J.M. & Kester, D.R. (Eds.), *Oceanic Processes in Marine Pollution, vol 1, Biological Processes and Wastes in the Ocean.* Krieger Publishing Co. Inc., Melbourne, pp. 37–49.

Buffagni, A., Erba, S., Birk, S., Cazzola, M., Feld, C., Ofenböck ,T., Murray-Bligh, J., Furse, M.T., Clarke, R., Hering, D., Soszka, H. & van de Bund, W. 2005. Towards European Inter-calibration for the Water Framework Directive: Procedures and examples for different river types from the E.C. project STAR.11th STAR deliverable. STAR Contract No: EVK1-CT 2001-00089. *Quad. Ist. Ric. Acque* 123, Rome (Italy), IRSA, 467 pp.

Buffagni, A., Erba, S., Cazzola, M., Murray-Bligh, J., Soszka, H. & Genoni, P. 2006. The STAR common metrics approach to the WFD intercalibration process: full application across Europe for small, lowland rivers. *Hydrobiologia*, 566: 379–399.

Buffagni, A., Erba, S. & Furse, M.T. 2007. A simple procedure to harmonize class boundaries of assessment systems at the pan-European scale. *Environmental Science and Policy*, 10: 709–724.

Buhr, K.J. 1976. Suspension feeding and assimilation efficency in *Lanice conchilega* (Polychaeta). *Marine Biology*, 38: 373–383.

Butcher, R.W. 1947. Studies in the ecology of rivers. IV. The algae of organically enriched water. *Journal of Ecology*, 35: 186–191.

Cabral, J.A., Pardal, M.A., Lopes, R.J., Múrias, T. & Marques, J.C. 1999. The impact of macroalgal blooms on the use of the intertidal area and feeding behaviour of waders (Charadrii) in the Mondego estuary (West Portugal). *Acta Oecologica*, 20(4): 417–428.

Caeiro, S., Costa, M.H., Ramos, T.B., Fernandes, F., Silveira, N., Coimbra, A., Medeiros, G. & Painho, M. 2005. Assessing heavy metal contamination in Sado Estuary sediment: an index analysis approach. *Ecological Indicators*, 5: 151–169.

Calderón-Aguilera, L.E. 1992. Análisis de la infauna béntica de Bahía de San Quintín, Baja California, con énfasis en su utilidad en la evaluación de impacto ambiental. *Ciencias Marinas*, 18(4): 27–46.

Calow, P. & Petts, G.E. 1992. *The Rivers Handbook: Hydrological and Ecological Principles.* Blackwell Publishing, Oxford, p. 536.

Canterbury, G.E., Martin, T.E., Petit, D.R., Petit, L.J. & Bradford, D.F. 2000. Bird communities and habitat as ecological indicators of forest condition in regional monitoring. *Conservation Biology*, 14: 544–558.

Cardoso, P.G., Pardal, M.A., Lillebø, A.I., Ferreira, S.M., Raffaelli, D. & Marques, J.C. 2004a. Dynamic changes in seagrass assemblages under eutrophication and implications for recovery. *Journal of Experimental Marine Biology and Ecology*, 302: 233–248.

Cardoso, P.G., Pardal, M.A., Raffaelli, D., Baeta, A. & Marques, J.C. 2004b. Macroinvertebrate response to different species of macroalgal mats and the role of disturbance history. *Journal of Experimental Marine Biology and Ecology*, 308: 207–220.

Carell, B., Forberg, S., Grundelius, E., Henrikson, L., Johnels, A., Lindh, U., Mutvei, H., Olsson, M., Svaerdstroem, K. & Westermark, T. 1987. Can mussel shells reveal environmental history? *Ambio*, 16(1): 2–10.

CARM. 1998. Ecosistemas de nuestra región. El medio marino. Agencia para el Medio Ambiente y la Naturaleza. Murcia.

Carmichael, J., Richardson, B., Roberts, M. & Jordan. S.J. 1992. *Fish sampling in eight Chesapeake Bay tributaries.* Maryland Department of Natural Resources, CBRM-HI-92-2. Annapolis, p. 50.

Carpene, E. 1993. Adaptations of marine organisms to heavy metal pollution: metallothioneins as biological markers. *Italian Journal of Biochemistry*, 42(5): 306.

Casselli, C., Ponti, M. & Abbiati, M. 2003. Valutazione della qualità ambientale della laguna costiera Pialassa Baiona attraverso lo studio dei suoi popolamenti bentonici. *XIII Congresso Societá Italiana de Ecología*, Como, Villa Olmo, 8–10 September 2003, poster.

Cemagref. 1982. Etude dês méthodes biologiques d'appréciation quantitative de la qualité dês eaux. Rapport Q.E, Lyon. Agence de l'Eau Rhone-Mediterranee-Corse-Cemagref. Lyon. France.

Chessman, B.C. 1995. Rapid assessment of rivers using macroinvertebrates: a procedure based on habitat-specific sampling, family level identification and a biotic index. *Australian Journal of Ecology*, 20: 122–129.

Chessman, B.C. 1997. Objective derivation of macroinvertebrate family sensitivity grade numbers for the SIGNAL biotic index: application to the Hunter River system, New South Wales. *Marine and Freshwater Research*, 48: 159–172.

Chutter, F.M. 1972. An empirical biotic index of the water quality in South African streams and rivers. *Water Research*, 6: 19–30.

Cicchetti, G., Latimer, J.S., Rego, S.A., Nelson, W.G., Bergen, B.J. & Coiro, L.L. 2006. Relationships between near-bottom dissolved oxygen and sediment profile camera measures. *Journal of Marine Systems*, 62: 124–141.

Clarke, K.R. 1990. Comparison of dominance curves. *Journal of Experimental Marine Biology and Ecology*, 138(1–2): 130–143.

Clarke, K.R., 1993. A method linking multivariate community structure to environmental variables. *Mar. Ecol. Progr. Ser.*, 92: 205–219.

Clarke, K.R. & Warwick, R.M. 1994. Change in marine communities: an approach to statistical analysis and interpretation. *Plymouth Marine Laboratory, Plymouth*, 144 p.

Clarke, K.R. & Warwick, R.M. 1998. A taxonomic distinctness index and its statistical properties. *Journal of Applied Ecology*, 35: 523–531.

Clarke, K.R. & Warwick, R.M. 1999. The taxonomic distinctness measures of biodiversity: weighting of steps lengths between hierarchical levels. *Marine Ecology-Progress Series*, 184: 21–29.

Clarke, K.R. & Warwick, R.M. 2001. A further biodiversity index applicable to species lists: variation in taxonomic distinctness. *Marine Ecology-Progress Series*, 216: 265–278.

Codling, I.D. & Ashley, S.J. 1992. Development of a biotic index for the assessment of pollution status of marine benthic communities. Final report to SNIFFER and NRA. NR 3102/1.

Cook, S.E. 1976. Quest for an index of community structure sensitive to water pollution. *Environmental Pollution*, 11: 269–288.

Cooper, J.A., Harrison, T.D., Ramm, A.E. & Singh, R.A. 1993. *Refinement, Enhancement and Application of the Estuarine Health Index to Natal'sEstuaries, Tugela – Mtamvuna.* Unpublished Technical Report. CSIR, Durban.

Cooper, R.J., Langlois, D. & Olley, J. 1982. Heavy metals in Tasmania shell fish. I. Monitoring heavy metals contamination in the Dermwert Estuary: use of oysters and mussels. *Journal of Applied Toxicology*, 2(2): 99–109.

Cormaci, M. & Furnari, G. 1991. Phytobenthic communities as monitor of the environmental conditions of the Brindisi coast-line. *Oebalia*, XVII-I(suppl.): 177–198.

Cossa, F., Picard, M. & Gouygou, J.P. 1983. Polynuclear aromatic hydrocarbons in mussels from the Estuary and northwestern Gulf of St. Lawrence, Canada. *Bulletin of Environmental Contamination and Toxicology*, 31: 41–47.

Cossa, A. & Rondeau, J.G. 1985. Seasonal geographical and size-induced variability in mercury content of *Mytilus edulis* in an estuarine environment: a re-assessment of mercury pollution level in the Estuary and Gulf of St. Lawrence. *Marine Biology*, 88: 43–49.

Costa, M.J., Almeida, P.R., Costa, J.L. & Assis, C.A. 1994. Do eel grass beds and salt marsh borders act as preferential nurseries and spawning grounds for fish? – an example of the Mira estuary in Portugal. *Ecological Engineering*, 3: 187–195.

Couch, J.A. & Harshbarger, J.C. 1985. Effects of carcinogenic agents on aquatic animals: an environmental and experimental overview. *Environmental Carcinogenesis Reviews*, 3C(1): 63–105.

Cox, E.J. 1991. What is the basis for using diatoms as monitors of river quality? In: Whitton, B.A., Rott, E. & Friedrich, G. (Eds.), *Use of Algae for Monitoring Rivers*. Institut für Botanik, Universität in Innsbruck, Innsbruck, pp. 33–40.

Craeymeersch, J.A. 1991. Applicability of the abundance/biomass comparison method to detect pollution effects on intertidal macrobenthic communities. *Hydrobiology Bulletin*, 24(2): 133–140.

Dabbas, M., Hubbard, F. & Mc Manus, J. 1984. The shell of *Mytilus* as an indicator of zonal variatons of water quality within an estuary. *Estuarine Coastal and Shelf Science*, 18(3): 263–270.

Dale, V.H. & Beyeler, S.C. 2001. Challenges in the development and use of ecological indicators. *Ecological Indicators*, 1: 3–10.

Dale, V.H., Beyeler, S.C. & Jackson, B. 2002. Understorey vegetation indicators of anthropogenic disturbance in longleaf pine forests at Fort Benning, Georgia, USA. *Ecological Indicators*, 1: 155–170.

Dauer, D.M. 1984. The use of polychaete feeding guilds as biological variables. *Marine Pollution Bulletin*, 15(8): 301–304.

Dauer, D.M. 1993. Biological criteria, environmental health and estuarine macrobenthic community structure. *Marine Pollution Bulletin*, 26: 249–257.

Dauer, D.M., Luckenbach, M.W. & Rodi, A.J., 1993. Abundance–biomass comparison ABC method: effects of an estuarine gradient, anoxic/hypoxic events and contaminated sediments. *Marine Biology*, 116: 507–518.

Dauer, D.M., Maybury, C.A. & Ewing, R.M. 1981. Feeding behavior and general ecology of several spionid solychaetes from the Chesapeake Bay. *Journal of Experimental Marine Biology and Ecology*, 54(1): 21–38.

Dauvin, J.C. & Ruellet, T. 2007. Polychaete/amphipod ratio revisited. *Marine Pollution Bulletin*, 55: 215–224.

De Boer, W.F., Daniels, P. & Essink, K. 2001. Towards ecological quality objectives for North Sea benthic Communities. National Institute for Coastal and Marine Management (RIKZ), Haren, the Netherlands. Contract RKZ 808, Report no 2001-11, 64 p.

Deegan, L.A., Finn, J.T., Ayvazian, S.G. & Ryder, C. 1993. *Feasibility and application of the Index of Biotic Integrity to Massachusetts Estuaries (EBI)*. Final report to Massachusetts Executive Office of Environmental Affairs. North Grafton, MA, Department of Environmental Protection.

DeKeyser, E.S., Kirby, D.R. & Ell, M.J. 2003. An index of plant community integrity: development of the methodology for assessing prairie wetland plant communities. *Ecological Indicators*, 3: 119–133.

Delong, M.D. & Brusven, M.A. 1998. Macroinvertebrate community structure along the longitudinal gradient of an agriculturally impacted stream. *Environmental Management*, 22: 445–457.

De Paw, N. & Vanhooren, G. 1983. Method for biological quality assessment of watercourses in Belgium. *Hydrobiologia*, 100: 153–168.

de Ruiter, P.C., Neutel, A.M. & Moore, J.C. 1994. Food webs and nutrient cycling in agro-ecosystems. *Trends in Ecology and Evolution*, 9/10: 378–383.

de Ruiter, P.C., Neutel, A.M. & Moore, J.C. 1995. Energetics, patterns of interaction strengths, and stability in real ecosystems. *Science*, 269: 1257–1260.

Descy, J.P. & Coste, M. 1990. Utilisation des diatomées benthiques pour l'évaluation de la qualité des eaux courrant. Rapport final. Université Namur, Cemagref Bordeaux CEE-Bm 112 p.

Descy, J.P. & Coste, M. 1991. A test of methods for assessing water quality based on diatoms. *Verh. Int. Verein. Limnol.*, 24: 2112–2116.

Devlin, M., Painting, S. & Best, M. 2007. Setting nutrient thresholds to support an ecological assessment based on nutrient enrichment, potential primary production and undisarable disturbance. *Marine Pollution Bulletin*, 55: 65–73.

De Wolf, P. 1975. Mercury content of mussel from West European Coasts. *Marine Pollution Bulletin*, 6(4): 61–63.

Docampo, L. & Bikuña, B. 1991. Analysis of the physico-chemical variables of the stream waters of Vizcaya (Basque Country): 1. Mathematical model of conductivity. *Archiv fuer Hydrobiologie*, 122(3): 351–372.

Dolbeth, M., Cardoso, P.G., Ferreira, S.M., Verdelhos, T., Raffaelli, D. & Pardal, M.A. 2007. Anthropogenic and natural disturbance effects on a macrobenthic estuarine community over a 10-year period. *Marine Pollution Bulletin*, 54: 576–585.

Dolbeth, M., Pardal, M.A., Lillebø, A.I., Azeiteiro, U. & Marques, J.C. 2003. Short- and long-term effects of eutrophication on the secondary production of an intertidal macrobenthic community. *Marine Biology*, 143: 1229–1238.

Drayton, B. & Primack, R.B. 1996. Plant species lost in an isolated conservation area in metropolitan Boston from 1894 to 1993. *Conservation Biology*, 10: 30–39.

Dzwonko, Z. & Loster, S. 1992. Species richness and seed dispersal to secondary woods in southern Poland. *Journal of Biogeography*, 19: 195–204.

Eadie, J.B., Faust, W., Gardner, W.S. & Nalepa, T. 1982. Polycyclic aromatic hydrocarbons in sediments and associated benthos in Lake Erie. *Chemosphere*, 11: 185–191.

EC, 2003. Common implementation strategy for the Water Framework Directive (2000/60/EC), Guidance Document No 5, Transitional and Coastal Waters – Typology, Reference Conditions and Classification Systems. European Communities, 107 pp.

EC, 2007a. WFD intercalibration technical report. Mediterranean GIG-Rivers. Benthic Macroinvertebrates, 17 p.

EC, 2007b. WFD intercalibration technical report. Mediterranean GIG-Rivers. Phytobenthos, 20 p.

Encalada, R.R. & Millan, E. 1990. Impacto de las aguas residuales industriales y domesticas sobre las comunidades bentónicas de la Bahía de Todos Santos, Baja California, Mexico. *Ciencias Marinas*, 16(4): 121–139.

Engle, V., Summers, J.K. & Gaston, G.R. 1994. A benthic index of environmental condition of Gulf of Mexico. *Estuaries*, 17(2): 372–384.

Engle, V.D. 2000. Application of the indicator evaluation guidelines to an index of Benthic condition for Gulf of Mexico Estuaries. In: Jackson, L.E., Kurtz, J.C. & Fisher, W.S. (Eds.), *Evaluation Guidelines for Ecological Indicators*. EPA/620/R-99/005. U.S.

Environmental Protection Agency, Office of Research and Development, Research Triangle Park, NC. 107 p.

Engle, V.D. & Summers, J.K. 1999. Refinement, validation, and application of a benthic condition index for Gulf of Mexico Estuaries. *Estuaries*, 22: 624–635.

EPA. 1987. *Biological Criteria for the Protection of the Aquatic Life*. Vols I–III. Surface Water Section, Division of Water Quality Monitoring and Assessment, Ohio Environmental Protection Agency, Columbus, OH.

FAME. 2004. Manual for the application of the European Fish Index – EFI. A fish-based method to assess the ecological status of European rivers in support of the Water Framework Directive. Version 1.1, January 2005.

Fano, E.A., Mistri, M. & Rossi, R. 2003. The ecofunctional quality index (EQI): a new tool for assessing lagoonal ecosystem impairment. *Estuarine Coastal and Shelf Science*, 56: 709–716.

Fasham, M.J.R. 1984. *Flows of energy and materials in marine ecosystems: theory and practice*. NATO Conf Ser IV. Marine science. Plenum, New York, p. 733.

Fauchald, K. & Jumars, P. 1979. The diet of the worms: a study of Polychaete feeding guilds. *Oceanography Marine Biology Annual Review*, 17: 193–284.

Feldmann, J., 1937. Recherches sur le vegetation marine de la Méditerranée. La côte des Albères. *Revue Algologique*, 10: 1–339.

Ferreira, J.G., Nobre, A.M., Simas, T.C., Silva, M.C., Newton, A., Bricker, S.B., Wolff, W.J., Stacey, P.E. & Sequeira, A. 2006. A methodology for defining homogeneous water bodies in estuaries – application to the transitional systems of the EU Water Framework Directive. *Estuarine, Coastal and Shelf Science*, 66: 468–482.

Finn, J.T. 1976 Measures of ecosystem structure and function derived from analyses of flows. *Journal of Theoretical Biology*, 41: 535–546.

Fisher, R.A., Corbet, A.S. & Williams, C.B. 1943. The relation between the number of species and the number of individuals in a random sample of an animal population. *Journal of Animal Ecology*, 12: 42–58.

Fonseca, J.C., Marques, J.C., Paiva, A.A., Freitas, A.M., Madeira, V.M.C. & Jørgensen, S.E. 2000. Nuclear DNA in the determination of weighing factors to estimate exergy from organisms biomass. *Ecological Modelling*, 126: 179–189.

Fonseca, J.C., Pardal, M.A., Azeiteiro, U.M. & Marques, J.C. 2002. Estimation of ecological exergy using weighing factors determined from DNA contents of organisms – a case study. *Hydrobiologia*, 475/476: 79–90.

Forni, G. & Occhipinti Ambrogi, A. 2003. Applicazione del coefficiente biotico (Borja *et al.*, 2000) alla comunità macrobentonica del Nord Adriático. *Meeting of the Italian Society of Marine Biology*, Tunisia, June 2003.

Frontier, S. & Pichod-Viale, D. 1995. *Ecosystèmes: Structure, Fonctionnement, Évolution*. 2nd edn. Masson, Paris, p. 447.

Fuliu, X. 1997. Exergy and estructural exergy as ecological indicators for the state of the Lake Chalou ecosystem. *Ecological Modelling*, 99: 41–49.

Fuller, R.J., Gough, S.J. & Marchant, J.H. 1995. Bird populations in new Lowland woods: landscape, design and management perspectives. In: Ferris-Kaan, R. (Ed.), *The Ecology of Woodland Creation*. Wiley, London, pp. 163–182.

Gambi, M.C. & Giangrande, A. 1985. Characterization and distribution of polychaete trophic groups in the soft-bottoms of the Gulf of Salerno. *Oebalia*, 11(1): 223–240.

Gamito, S., Gilabert, J., Marcos, C. & Pérez-Ruzafa, A. 2005. Effects of changing environmental conditions on lagoon ecology. In: Gönenc, J.E. & Wolflin, J. (Eds.), *Coastal Lagoons: Ecosystem Processes and Modelling for Sustainable Use and Development*. CRC Press, Boca Raton, pp. 193–229.

Gauch, H.G. 1982. *Multivariate Analysis in Community Ecology. Cambridge Studies in Ecology.* Cambridge University Press, New York.

Gee, J.M., Warwick, R.M., Schaanning, M., Berge, J.A. & Ambrose, W.G. 1985. Effects of organic enrichment on meiofaunal abundance and community structure in sublittoral soft sediments. *Journal of Experimental Marine Biology and Ecology*, 91(3): 247–262.

Gibbs, P.E., Langston, W.J., Burt, G.R. & Pascoe, P.L. 1983. *Tharyx marioni* (Polychaeta): a remarkable accumulator of arsenic. *Journal of the Marine Biological Association of the United Kingdom*, 63(2): 313–325.

Gibson, G.R., Bowman, M.L., Gerritsen, J. & Snyder, B.D. 2000. Estuarine and Coastal Marine Waters: Bioassessment and Biocriteria Technical Guidance. EPA 822–B–00–024. U.S. Environmental Protection Agency, Office of Water, Washington, DC.

Giovanardi, F. & Vollenweider, R.A. 2004. Trophic conditions of marine coastal waters: experience in applying the Trophic Index TRIX to two areas of the Adriatic and Tyrrhenian seas. *Journal of Limnology*, 63: 199–218.

Glémarec, M., 1986. Ecological impact of an oil-spill: utilization of biological indicators. IAWPRC-NERC Conference, July 1985. *IAWPRC Journal*, 18: 203–211.

Glémarec, M. & Hily, C. 1981. Perturbations apportées a la macrofaune benthique de la Baie de Concarneau par les effluents urbains et portuaires. *Acta Oecology Applications*, 2(2): 139–150.

Goff, F.G. & Cottam, G. 1967. Gradient analysis: The use of species and symthetic indices. *Ecology*, 48: 793–806.

Goldberg, E., Bowen, V.T., Farrington, J.W., Harvey, G., Martin, P.L., Parker, P.L., Risebrough, R.W., Robertson, W., Schnider, E. & Gamble, E. 1978. The Mussel Watch. *Environmental Conservation*, 5: 101–125.

Golovenko, V.K., Shchepinsky, A.A. & Shevchenko, V.A. 1981. Accumulation of DDT and its metabolism in Black Sea Mussels. Izvestiia Akademii nauk Gruzinskoï SSR. Seriia biologicheskaia, 4: 453–550.

Gómez-Gesteira, J.L. & Dauvin, J.C. 2000. Amphipods are good bioindicators of the impact of oil spillson soft bottom macrobenthic communities. *Marine Pollution Bulletin*, 40(11): 1017–1027.

Gorostiaga, J.M., Borja, Á., Díez, I., Francés, G., Pagola-Carte, S. & Sáiz-Salinas, J.I. 2004. Recovery of benthic communities in polluted systems. In: Borja, Á. & Collins, M. (Eds.), *Oceanography and Marine Environment of the Basque Country, Elsevier Oceanography Series*, Elsevier, Amsterdam, Vol. 70, pp. 549–578.

Gosset, R.W., Brown, D.A. & Young, D.R. 1983. Predicting the bioaccumulation of organic compounds in marine organisms using octanol/water partition coefficients. *Marine Pollution Bulletin*, 14: 387–392.

Grall, J. & Glémarec, M., 1997. Using biotic indices to estimate macrobenthic community perturbations in the Bay of Brest. *Estuarine, Coastal and Shelf Science*, 44(Suppl. A): 43–53.

Gray, J.S. 1979. Pollution-induced changes in populations. *Philosophical Transactions of the Royal Society London*, 286: 545–561.

Gray, J.S., Aschan, M., Carr, M.R., Clarke, K.R., Green, R.H., Pearson, T.H., Rosenberg, R., Warwick, R.M. & Bayne, B.L. 1988. Analysis of community attributes of the benthic macrofauna of Frierfjord/Langesundfjord and in a mesocosm experiment. In: Warwick, R.M. & Clarke, K.R. (Eds.), *Biological Effects of Pollutants. Results of a Practical Workshop*, 46(1–3): 151–165.

Gray, J.S. & Mirza, F.B. 1979. A possible method for the detection of pollution induced disturbance on marine benthic communities. *Marine Pollution Bulletin*, 10: 142–146.

Gray, J.S. & Pearson, T.H. 1982. Objective selection of sensitive species indicative of pollution-induced change in benthic communities: A comparative methodology. *Marine Ecology Progress Series*, 9: 111–119.

Guinda, X., Juanes, J.A., Puente, A. & Revilla, J.A. 2008. Comparison of two methods for quality assessment of macroalgae assemblages, under different pollution types. *Ecological Indicators*, 8(5): 743–753.

Hale, S.S. & Heltshe, J.F. 2008. Signals from the benthos: development and evaluation of a benthic index for the nearshore Gulf of Maine. *Ecological Indicators*, 8: 338–350.

Hall, S.J. & Greenstreet, S.P. 1998. Taxonomic distinctness and diversity measures: responses in marine fish communities. *Marine Ecology Progress Series*, 166: 227–229.

Hammond, P.C. & Miller, J.C. 1998. Comparison of the biodiversity of Lepidoptera within three forested ecosystems. *Annals of the Entomological Society of America*, 91: 323–328.

Hansen, P.D. & Addison, R.F. 1990. The use of mixed function oxidases (MFO) to support biological effects monitoring in the sea. *Copenhagen Denmark Ices.*, 1: 1–9.

Harrison, F.L., Lam, J.R. & Novacek, J. 1988. Partitioning of metals among metal-binding proteins in the bay mussel, *Mytilus edulis*. *Response of Marine Organisms to Pollutants*, 24(1–4): 167–170.

Harrison, T.D., Cooper, J.A., Ramm, A.E. & Singh, R.A. 1994. *Application of the Estuarine Health Index to South Africa's estuaries, Orange River – Buffels (Oos)*. Unpublished Technical Report. CSIR, Durban.

Harrison, T.D., Cooper, J.A., Ramm, A.E. & Singh, R.A. 1995. *Application of the Estuarine Health Index to South Africa's estuaries, Palmiet – Sout*. Unpublished Technical Report. CSIR, Durban.

Harrison, T.D., Cooper, J.A., Ramm, A.E. & Singh, R.A. 1997. *Application of the Estuarine Health Index to South Africa's estuaries, Old Woman's – Great Kei*. Unpublished Technical Report. CSIR, Durban.

Harrison, T.D., Cooper, J.A., Ramm, A.E. & Singh, R.A. 1999. *Application of the Estuarine Health Index to South Africa's estuaries, Transkei*. Unpublished Technical Report. CSIR, Durba.

Heino, J., Soininen, J., Lappalainen, J. & Virtanen, R. 2005. The relationship between species richness and taxonomic distinctness in freshwater organisms. *Limnology and Oceanography*, 50(3), 978–986.

Heip, C., Vincx, M. & Vraken, G. 1985. The ecology of marine nematodes. *Oceanography and Marine Biology: An Annual Review*, 23: 399–489.

Hellawell, V.M. 1986. *Biological Monitors of Freshwater Pollution and Environmental Management*. Elsevier, London.

Hilsenhoff, W.L. 1987. An improved biotic index of organic stream pollution. *The Great Lakes Entomologist*, 20: 31–39.

Hilsenhoff, W.L. 1988. Rapid field assessment of organic pollution with a family-level biotic index. *Journal of the North American Benthological Society*, 7: 65–68.

Hily, C., 1984. Variabilité de la macrofauna benthique dans les milieux hypertrophiques de la Rade de Brest. Thése de Doctorat d'Etat, Univ. Bretagne Occidentale. Vol. 1, 359 pp; Vol. 2, 337 pp.

Hobbs, R.J. 1997. Can we use plant functional types to describe and predict responses to environmental change? In: Smith, T.M., Shugart, H.H. & Woodward, F.I. (Eds.), *Plant Functional Types Their Relevance to Ecosystem Properties and Global Change*. Cambridge University Press, Cambridge, pp. 66–90.

Holl, K.D. 1996. The effect of coal surface mine reclamation on diurnal lepidopteran conservation. *Journal of Applied Ecology*, 33: 225–236.

Holliday, N.J. 1991. Species responses of carabid berttles (Colepoptera: Carabidae) during post-fire regeneration of boreal forest. *The Canadian Entomologist,* 123: 1369–1389.

Hong, J. 1983. Impact of the pollution on the benthic community. *Bulletin of Korean Fisheries Society*, 16: 273–290.

Hughes, R.G. 1984. A model of the structure and dynamics of benthic marine invertebrate communities. *Marine Ecology Progress Series*, 15(1–2): 1–11.

Hulbert, S.H. 1971. The nonconcept of species diversity: a critique and alternative parameters. *Ecology*, 52: 577–586.

Ibanez, F. & Dauvin, J.C. 1988. Long-term changes (1977–1987) in a muddy fine sand *Abra alba–Melina palmata* community from the Western Channel: multivariate time series analysis. *Marine Ecology Progress Series*, 19: 65–81.

ISO. 1984. Water assessment of the water and habitat quality of rivers by a macroinvertebrate "score". ISO/TC 147/SC 5/ WG 6 N 4, Draft proposal ISO/ DP 8689.

Jeffrey, D.W., Wilson, J.G., Harris, C.R. & Tomlinson, D.L. 1985. The application of two simple indices to Irish estuary pollution status. In: Wilson, J.G. & Halcrow, W. (Eds.), *Estuarine Management and Quality Assessment*. Plenum Press, London.

Jensen, K., Randlov, A. & Riisgard, H.U. 1981. Heavy metal pollution from a point source demonstrated by means of mussels, *Mytilus edulis*. *Chemosphere*, 10(7): 761–765.

Jensen, P. 1987. Feeding ecology of free-living aquatic nematodes. *Marine Ecology Progress Series*, 35: 187–196.

Jordan, S.J., Vaas, P.A. & Uphoff, J. 1990. Fish assemblages as indicators of environmental quality in Chesapeake Bay. In: *Biological Criteria: Research and Regulation*, December 12–13, 1990, Arlington, VA. U.S. Environmental Protection Agency, Office of Water.

Jørgensen, S.E. 1994. Review and comparison of goal functions in system ecology. *Vie Milieu*, 44(1): 11–20.

Jørgensen, S.E. & Bendoricchio, G. 2001. *Fundamentals of Ecological Modelling*, Elsevier, Oxford, p. 530.

Jørgensen, S.E., Fath, B., Bastianoni, S., Marques, J.C., Müller, F., Nielsen, S.N., Patten, B.C., Tiezzi, E. & Ulanowicz, R.E. 2007. *A New Ecology. Systems Perspective*. Elsevier, p. 275.

Jørgensen, S.E., Ladegaard, N., Debeljak, M. & Marques, J.C. 2005. Calculations of exergy for organisms. *Ecological Modelling*, 185: 165–175.

Jørgensen, S.E. & Marques, J.C. 2001. Thermodynamics and ecosystem theory. Case studies from Hydrobiology. *Hydrobiologia*, 445: 1–10.

Jørgensen, S.E., Marques, J.C. & Nielsen, S.N. 2002. Structural changes in an estuary, described by models and using Exergy as orientor. *Ecological Modelling*, 158: 233–240.

Jørgensen, S.E. & Mejer, H. 1979. A holistic approach to ecological modelling. *Ecological Modelling*, 7: 169–189.

Jørgensen, S.E. & Mejer, H. 1981. Exergy as a key function in ecological models. In: Mitsch, W., Bosserman, R.W. & Klopatek, J.M. (Eds.), *Energy and Ecological Modelling. Developments in Environmental Modelling*, Elsevier. Amsterdam, Vol. 1, pp. 587–590.

Jørgensen, S.E., Nielsen, S.N. & Mejer, H. 1995. Energy, environ, exergy and ecological modelling. *Ecological Modelling*, 77: 99–109.

Jørgensen, S.E. & Padisak, J. 1996. Does the intermediate disturbance hypothesis comply with thermodynamics? *Hydrobiologia*, 323: 9–21.

Juanes, J.A., Guinda, X., Puente, A. & Revilla, J.A. 2008. Macroalgae, a suitable indicator of the ecological status of coastal rocky communities in the NE Atlantic. *Ecological Indicators*, 8: 351–359.

Karr, J.R. 1981. Assessment of biotic integrity using fish communities. *Fisheries*, 6(6): 21–27.

Kelly, M.G., Adams, C., Graves, A.C., Jamieson, J., Krokowski, J., Lycett, E.B., Murray-Bligh, J., Pritchard, S. & Wilkins, C. 2001. The Trophic Diatom Index: A User's Manual. Revised Edition. R&D Technical Report E2/TR2, Bristol: Environment Agency.

Kelly, M.G. & Whitton, B. 1995. The trophic diatom index: a new index for monitoring eutrophication in rivers. *Journal of Applied Phycology*, 7: 433–444.

Kimberling, D., Karr, J.R. & Fore, L.S. 2001. Measuring human disturbance using terrestrial in vertebrates in the shrub-steppe of eastern Washington (USA). *Ecological Indicators*, 1: 63–81.

Kolkwitz, R. & Marsson, M. 1908. Ökologie der pflanzlichen Saprobien. *Berichte der Deutschen Botanischen Gesellschaft*, 26(A): 505–519.

Koskimies, P. 1989. Birds as a tool in environmental monitoring. *Annales Zoologici Fennici*, 6: 153–166.

Krause-Jensen, D., Carstensen, J. & Dahl, K. 2007. Total and opportunistic algal cover in relation to environmental variables. *Marine Pollution Bulletin*, 55: 114–125.

Kryazheva, N.G., Chistyakova, E.K. & Zakharov, V.M. 1996. Analysis of development stability of *Betula pendula* under conditions of chemical pollution. *Russian Journal of Ecology*, 27: 422–424.

Kurelec, B., Kezic, N., Singh, H. & Zahn, R.K. 1984. Mixed-function oxidases in fish: Their role in adaptation to pollution. *Responses of Marine Organisms to Pollutants*, 14(1–4): 409–411.

Labrune, C., Amouroux, J.M., Sarda, R., Dutrieux, E., Thorin, S., Rosenberg, R. & Grémare, A. 2006. Characterization of the ecological quality of the coastal Gulf of Lions (NW Mediterranean). A comparative approach based on three biotic indices. *Marine Pollution Bulletin*, 52: 34 – 47.

Lafaurie, M., Mathieu, A., Salaun, J.P., Narbonne, J.F., Galgani, F., Romeo, M., Monod, J.L. & Garrigues, P.H. 1993. Biochemical markers in *pollution assessment*: field studies along the North coast of the Mediterranean Sea. *Mapping and Technical Reports Series*, 69: 1–123.

Lambshead, P.J. & Platt, H.M. 1985. Structural patterns of marine benthic assemblages and their relationship with empirical statistical models. In: Gibbs, P.E. (Ed.), *Proceedings of the Nineteenth European Marine Biology Symposium*: Plymouth, pp. 16–21.

Lambshead, P.J., Platt, H.M. & Shaw, K.M. 1983. The detection of differences among assemblages of marine benthic species based on an assessment of dominance and diversity. *Journal of Natural History*, 17: 859–847.

Langston, W.J., Burt, G.R. & Zhou, M.J. 1987. Tin and organotin in water, sediments and benthic organisms of Poole Harbour. *Marine Pollution Bulletin*, 18(12): 634–639.

Lauenstein, G., Robertson, A. & O'Connor, T. 1990. Comparison of trace metal data in mussels and oysters from a Mussel Watch programme of the 1970s with those from a 1980s programme. *Marine Pollution Bulletin*, 21(9): 440–447.

Leary, R.F. & Allendorf, F.W. 1989. Fluctuating asymmetry as an indicator of stress in conservation biology. *Trends in Ecology and Evolution*, 4: 214–217.

Leonard, D.R., Clarke, K.R., Somerfield, P.R. & Warnick, R.M. 2006. The application of an indicator based on taxonomic distinctness for UK marine biodiversity assessment. *Journal of Enviromental Managment*, 78: 52–62.

Leonzio, C., Bacci, E., Focardi, S. & Renzoni, A. 1981. Heavy metals in organisms from the northern Tyrhenian sea. *Science of The Total Environment*, 20(2): 131–146.

Leppäkoski, E. 1975. Assessment of degree of pollution on the basis of macrozoobenthos in marine and brakish water environments. *Acta Academiae Aboensis*, 35: 1–98.

Levine, H.G. 1984.The use of seaweeds for monitoring coastal waters. In: Shubert, L.E. (Ed.), *Algae as Ecological Indicators*. Academic Press, London, p. 434.

Lewin, B. 1994. *Genes*. V. Oxford University Press, Oxford, p. 620.

Lillebø, A.I., Neto, J.M., Martins, I., Verdelhos, T., Leston, S., Cardoso, P.G., Ferreira, S., Marques, J.C. & Pardal, M.A. 2005. Management of a shallow temperate estuary to control eutrophication: the effect of hydrodynamics on the system's nutrient loading. *Estuarine, Coastal and Shelf Science*, 65: 697–707.

Lillebø, A.I., Teixeira, H., Pardal, M.A. & Marques, J.C. 2007. Applying quality status criteria to a temperate estuary before and after the mitigation measures to reduce eutrophication symptoms. *Estuarine, Coastal and Shelf Science*, 72: 177–187.

Llansó, R.J., Scott, L.C., Dauer, D.M., Hyland, J.L. & Russell, D.E. 2002a. An estuarine Benthic Index of biotic integrity for the mid-Atlantic region of the United States. I: Classification of assemblages and habitat definition. *Estuaries*, 25(6A): 1219–1230.

Llansó, R.J., Scott, L.C., Hyland, J.L., Dauer, D.M., Russell, D.E. & Kurtz, F.W. 2002b. An estuarine Benthic Index of biotic integrity for the mid-Atlantic region of the United States. II: Index development. *Estuaries*, 25(6A): 1231–1242.

Lopes, R.J., Múrias, T., Cabral, J.A. & Marques, J.C. 2005. A ten year study of variation, trends and seasonality of shorebird community in the Mondego estuary, Portugal. *Waterbirds*, 28: 8–18.

Lopes, R.J., Pardal, M.A., Múrias, T., Cabral, J.A. & Marques, J.C. 2006. Influence of macroalgal mats on abundance and distribution of dunlin *Calidris alpina* in estuaries: a long-term approach. *Marine Ecology Progress Series*, 323: 11–20.

Losovskaya, G.V. 1983. On significance of polychaetes as posible indicators of the Black Sea environment quality. *Ekologiya Moray Publications*, 12: 73–78.

Louette, M., Bijnens, L., Upoki Agenong', A.D. & Fotso, R.C. 1995. The utility of birds as bioindicators: case-studies in Equatorial Africa. *Belgian Journal of Zoology*, 125: 157–165.

Loureiro, S., Newton, A. & Icely, J. 2006. Boundary conditions for the European Water Framework Directive in the Ria Formosa lagoon, Portugal (physico-chemical and phytoplankton quality elements). *Estuarine, Coastal and Shelf Science*, 67: 382–398.

Lowe, R.L. & Pan, Y. 1996. Benthic algal communities and biological monitors. In: Stevenson, R.J., Bothwell, M. & Lowe, R.L. (Eds.), *Algal Ecology: Freshwater Benthic Ecosystems*. Academic Press, San Diego, California, USA, pp. 705–739.

MacArthur, R.H. 1972. *Geographical Ecology. Patterns in the Distribution of Species.* Harper and Row, New York.

Macauley, J.M., Summers, J.K. & Engle, V.D. 1999. *Estimating the Ecological Condition of the estuaries of the Gulf of Mexico.* Environmental Monitoring and Assessment.

Maeda, S. & Sakaguchi, T. 1990. Accumulation and detoxification of toxic metal elements by algae. In: Akatsuka, I. (Ed.), *Introduction to Applied Phycology*. SPB Academic Publishing, The Hague, The Netherlands, pp. 109–136.

Magurran, A.E. 1989. *Diversidad ecológica y su medición*. Vedrá. Barcelona.

Maine Department of Environmental Protection. 1987. *Methods for Biological Sampling and Analysis of Maines's Waters*. Maine Department of Environmental Protection, Augusta, ME.

Majer, J.D., Day, J.E., Kabay, E.D. & Perriman, W.S. 1984. Recolonization by ants in bauxite mines rehabilitated by a number of different methods. *Journal of Applied Ecology*, 21: 355–375.

Mal, T.K., Uveges, J.L. & Turk, K.W. 2002. Fluctuating asymmetry as an ecological indicator. *Ecological Indicators*, 1: 189–195

Malins, D.C., Mccain, B.B., Myers, M.S., Brown, D.W., Sparks, A.K., Morado, J.F. & Hodgins, H.O. 1984. Toxic chemicals and abnormalities in fish and shellfish from urban bays of Puget Sound. *Responses of Marine Organisms to Pollutants*, 14(1–4): 527–528.

Margalef, R. 1969. *Perspectives in Ecological Theory*. The University of Chicago Press, Chicago, p. 111.

Margalef, R. 1978. *Ecología*. Omega, Barcelona.

Margalef, R. 1983. *Limnología*. Omega, Barcelona.

Margalef, R. 1989. Reflexiones sobre la diversidad y significado de su expresión cuantitativa. *Simposio sobre la Diversidad Biológica*: 1–16. Madrid.

Margalef, R. 1991. *Teoría de los sistemas ecológicos*. Publications de la Universitat de Barcelona, Barcelona, p. 290.

Marina, M. & Enzo, O. 1983. Variability of zinc and manganeso concentrations in relation to sex and season in the bivalve *Donax trunculus*. *Marine Pollution Bulletin*, 14(9): 342–346.

Marín-Guirao, L., Cesar, A., Marín, A., Lloret, J. & Vita, R. 2005. Establishing the ecological quality status of soft-bottom mining-impacted coastal water bodies in the scope of the Water Framework Directive. *Marine Pollution Bulletin*, 50(4): 374–387.

Marques, J.C. 2001. Diversity, biodiversity, conservation, and sustainability. *The Scientific World*, 1: 534–543.

Marques, J.C., Maranhão, P. & Pardal, M.A. 1993. Human impact assessment on the subtidal macrobenthic community structure in the Mondego estuary (Western Portugal). *Estuarine Coastal and Shelf Science*, 37: 403–419.

Marques, J.C., Neto, J.M., Patrício, J., Pinto, R., Teixeira, H. & Veríssimo, H. 2007. Monitoring the Mondego estuary. Anthropogenic changes and their impact on ecological quality. Preliminary results from the first assessment of the effects of reopening the communication between the North and South arms on the eutrophication state of the system. Final Report, January 2007. IMAR/INAG, 87 pp.

Marques, J.C., Nielsen, S.N., Pardal, M.A. & Jørgensen, S.E. 2003. Impact of eutrophication and river management within a framework of ecosystem theories. *Ecological Modelling*, 166: 147–168.

Marques, J.C., Pardal, M.A., Nielsen, S.N. & Jørgensen, S.E. 1997. Analysis of the properties of exergy and biodiversity along an estuarine gradient of eutrophication. *Ecological Modelling*, 102: 155–167.

Marques, J.C., Pardal, M.A., Nielsen, S.N. & Jørgensen, S.E. 1998. Thermodynamic orientors: exergy as a holistic ecosystem indicator: a case study. In: Müller, F. & Leupelt, M. (Eds.), *Ecotargets, Goal Functions and Orientors. Theorical Concepts and Interdisciplinary Fundamentals for an Integrated, System Based Environmental Management*. Springer-Verlag, Berlin, pp. 87–101.

Marques, J.C., Teixeira, H., Patrício, J. & Neto, J.M. 2005. Avaliação do impacto das obras de interrupção da ligação entre os dois braços do estuário do Mondego na qualidade ecológica do sistema. Propostas de solução. IMAR, Relatório Técnico, 99 pp.

Martins, I., Lopes, R.J., Lillebø, A.I., Neto, J.M., Pardal, M.A., Ferreira, J.G. & Marques, J.C. 2007. Significant variations in the productivity of green macroalgae in a mesotidal estuary: implications to the nutrient loading of the system and the adjacent coastal area. *Marine Pollution Bulletin*, 54: 678–690.

Martins, I., Neto, J.M., Fontes, M.G., Marques, J.C. & Pardal, M.A. 2005. Seasonal variation in short-term survival of *Zostera noltii* transplants in a declining meadow in Portugal. *Aquatic Botany*, 82: 132–142.

Martins, I., Oliveira, J.M., Flindt, M.R. & Marques, J.C. 1999. The effect of salinity on the growth rate of the macroalgae *Enteromorpha intestinalis* (Chlorophyta) in the Mondego estuary (west Portugal). *Acta Oecologic*, 20: 259–265.

Martins, I., Pardal, M.A., Lillebø, A.I., Flindt, M.R. & Marques, J.C. 2001. Hydrodynamics as a major factor controlling the occurrence of green macroalgae blooms in an eutrophic estuary: a case study. *Estuarine Coastal and Shelf Science*, 52: 165–177.

Mason, D. 1996. Responses of Venezuelan uderstory birds to selective logging, enrichment strips, and vine cutting. *Biotroptica*, 28: 296–309.

Matlack, G.R. 1994. Plant species migration in a mixed-history forest landscape in Eastern North America. *Ecology*, 75: 1491–1502.

Maurer, D. & Leathem, W. 1981. Polychaete feeding guilds from Georges Bank, USA. *Marine Biology*, 62(2–3): 161–171.

Maurer, D., Leathem, W. & Menzie, C. 1981. The impact of drilling fluid and well cuttings on polychaete feeding guilds from the US Northeastern Continental Shelf. *Marine Pollution Bulletin*, 12(10): 342–347.

May, R.M. 1975. Patterns of species abundances and diversity. In: Cody, M.L. & Diamond, J.M. (Eds.), *Ecology and Evolution of Communities*. Belknap Press of Harvard University Press, Cambridge, MA, pp. 81–120.

McArdle, B.H. & Anderson, M.J. 2001. Fitting multivariate models to community data: a comment on distance based redundancy analysis. *Ecology*, 82: 290–297.

McElroy, A.E. 1988. Trophic transfer of PAH and metabolites (fish, worm). *Responses of Marine Organisms to Pollutants*, 8(1–4): 265–269.

McGeoch, M.A. 1998. The selections, testing and application of terrestrial insects as bioindicators. *Biological Reviews*, 73: 181–201.

McGinty, M. & Linder, C. 1997. An estuarine index of biotic integrity for Chesapeake Bay tidal fish community. In: Hartwell, S. (Ed.), *Biological Habitat Quality Indicators for Essential Fish Habitat.* NOAA Technical Memorandum NMFS-F/SPO-32, p. 124.

McIntyre, S., Lavorel, S. & Tremont, R.M. 1995. Plant life-history attributes: their relationship to disturbance in herbaceous vegetation. *Journal of Ecology*, 83: 31–44.

McIver, J.D., Parsons, G.L. & Moldenke, A.R. 1992. Litter spider succession after clear-cutting in a western coniferous forest. *Canadian Journal of Forestry Research*, 22: 984–992.

McLachlan, S.M. & Bazely, D.R. 2001. Recovery patterns of understorey herbs and their use as indicators of deciduous forest regeneration. *Conservation Biology*, 15: 98–110.

McManus, J.W. & Pauly, D. 1990. Measuring ecological stress: Variations on a theme by R.M. Warwick. *Marine Biology*, 106 (2): 305–308.

Meire, P.M. & Dereu, J. 1990. Use of the abundance/biomass comparison method for detecting environmental stress: some considerations based on intertidal macrozoobenthos and bird communities. *Journal of Applied Ecology*, 27(1): 210–223.

Mendez-Ubach, N. 1997. Polychaetes inhabiting soft bottoms subjected to organic enrichment in the Topolobampo lagoon complex, Sinaloa, México. *Océanides*, 12(2): 79–88.

Miller, B.S. 1986. Trace metals in the common mussel *Mytilus edulis* (L.) in the Clyde Estuary. *Proceedings of the Royal Society of Edinburgh*, 90: 379–391.

Milovidova, N.Y. & Alyomov, S.V. 1992. Physical/chemical characteristics of bottom sediments and soft bottom zoobenthos of Sevastopol region. In: Polykatpov, G. (Ed.), *Molysmology of the Black Sea*. Naukova Dumka, Kiev, pp. 216–282.

Mo, C. & Neilson, B. 1991. Variability in measurements of zinc in oysters, *C. virginica*. *Marine Pollution Bulletin*, 22(10): 522–525.

Moens, T., van Gansbeke, D. & Vincx, M. 1999a. Linking estuarine nematodes to their suspected food. A cause study from the Westerschelde estuary (south-west Netherlands). *Journal of the Marine Biological Association of the United Kingdom*, 79: 1017–1027.

Moens, T. & Vincx, M. 1997. Observations on the feeding ecology of estuarine nematodes. *Journal of the Marine Biological Association of the United Kingdom*, 77: 211–227.

Moffatt, S.F. & McLachlan, S.M. 2004. Understorey indicators of disturbance for riparian forests along an urban–rural gradient in Manitoba. *Ecological Indicators*, 4: 1–16.

Mohlenberg, F. & Riisgard, H.U. 1988. Partioning of inorganic and organic mercury in cockles *Cardium edule* (L.) and *C. glaucum* (Bruguiere) from a chronically polluted area: influence of size and age. *Environmental Pollution*, 55: 137–148.

Möller, A.P. 1998. Developmental instability of plants and radiation from Chernobyl. *Oikos*, 81: 444–448.

Möller, A.P. & Swaddle, J.P. 1997. *Asymmetry, Developmental Stability, and Evolution*. Oxford University Press, Oxford.

Molvær, J., Knutzen, J., Magnusson, J., Rygg, B., Skei, J. & Sørensen, J. 1997. *Classification of Environmental Quality in Fjords and Coastal Waters*. SFT guidelines. 97(03): 36 p.

Monaco, M.E. & Ulanowicz, R.E. 1997. Comparative ecosystem trophic structure of three U.S. mid-Atlantic estuaries. *Marine Ecology Progress Series*, 161: 239–254.

Moore, J.C. & de Ruiter, P.C. 1991 Temporal and spatial heterogeneity of trophic interactions within belowground food webs. *Agriculture, Ecosystems and Environment*, 34: 371–397.

Moore, J.C. & de Ruiter, P.C. 1997a. A food web approach to disturbance and ecosystem stability. In: van Straalen, N.M. & Løkke, H. (Eds.), *Ecological Principles for Risk Assessment of Contaminants in Soil*. Chapman & Hall, Dordrecht Netherlands.

Moore, J.C. & de Ruiter, P.C. 1997b. Resource compartmentation and the stability of real ecosystems. In: Gange, A.C. & Brown, V. (Eds.), *Multitrophic Interactions in Terrestrial Ecosystems*. Blackwell Scientific Publications, Oxford.

Moore, J.C., de Ruiter, P.C. & Hunt, H.W. 1993. Influence of productivity on the stability of real and model ecosystems. *Science*, 261: 906–908.

Moreno, D., Aguilera, P.A. & Castro, H. 2001. Assessment of the conservation status of seagrass (*Posidonia oceanica*) meadows: implications for monitoring strategy and the decision-making process. *Biological Conservation*, 102: 325–332.

Morrison, M.L. 1986. Bird populations as indicators of environmental changes. *Current Ornithology*, 3: 429–451.

Muniz, P., Venturini, N., Pires-Vanin, A.M.S., Tommasi, L.R. & Borja, A. 2005. Testing the applicability of a Marine Biotic Index (AMBI) to assessing the ecological quality of soft-bottom benthic communities, South America Atlantic region. *Marine Pollution Bulletin*, 50: 624–637.

Muxika, I., Borja, A. & Bald, J. 2007. Using historical data, expert judgement and multivariate analysis in assessing reference conditions and benthic ecological status, according to the European Water Framework Directive. *Marine Pollution Bulletin*, 55(1–6): 16–29.

Muxika, I., Borja, Á. & Bonne, W. 2005. The suitability of the marine biotic index (AMBI) to new impact sources along European coasts. *Ecological Indicators*, 5: 19–31.

Muxika, I., Borja, Á. & Franco, J. 2003. The use of Biotic Index (AMBI) to identify spatial and temporal gradients on benthic communities in an estuarine area. ICES CM 2003/ Session J-01. Tallinn (Estonia), 24–28 September, 2003.

Myers, M.S., Rhodes, L.D. & McCain, B.B. 1987. Pathologic anatomy and patterns of occurrence of hepatic neoplasm, putative preneoplastic lesions and other idiopathic hepatic conditions in English sole (*Parophrys vetulus*) from Puget Sound, Washington, U.S.A. *Journal of the National Cancer Institute*, 78(2): 333–363.

Nash, M.S., Whitford, W.G., Van Zee, J. & Havstad, K. 1998. Monitoring changes in stressed ecosystems using spatial patterns of ant communities. *Environmental Monitoring and Assessment*, 51: 201–210.

NEAGIG. 2004. Minutes of Northeast Atlantic Geographical Intercalibration Group Benthic Expert Meeting. Kristineberg Marine Station, Fiskebäckskil, Sweden, 22–24 September 2004.

Nelson, W.G. 1990. Prospects for development of an index of biotic integrity for evaluating habitat degradation in coastal systems. *Chemical Ecology*, 4: 197–210.

Neto, J.M. 2004. *Nutrient enrichment in a temperate macro-tidal system. Scenario analysis and contribution to coastal management.* Ph.D. Thesis. University of Coimbra, Portugal, 139 pp.

Neto, J.M., Flindt, M.R., Marques, J.C. & Pardal, M.A. 2008. Modelling nutrient mass balance in a temperate macro-tidal estuary: implications to management. *Estuarine, Coastal and Shelf Science*, 76: 175–185.

Neto, J.M., Gaspar, R., Pereira, L. & Marques, J.C. Ecological quality assessment of intertidal rocky shores. The use of marine macroalgae under the scope of European Water Framework Directive. *Ecological Indicators.* (submitted)

Neumann, G. 1992. Responses of foraging ant population to high intensity wildfire, salvage logging, and natural regeneration processes in *Eucalyptus regnans* regrowth forest of the Victorian central highlands. *Australian Forestry*, 55: 29–38.

Neumann, G., Notter, M. & Dahlgaard, H. 1991. Bladder-wrack (*Fucus vesiculosus* L) as an indicator for radionclides in the environment of Swedish nuclear power plants. *Swedish Environmental Protection Agency*, 3931: 1–35.

Nicholson, S. & Hui, Y.H. 2003. *Ecological Monitoring for Uncontaminated Mud Disposal Investigation. First Monitoring Report, East of Ninepins.* Civil Engineering Department, Fill Management Division, The Government of Hong Kong Special Administrative Region, Mouchel Asia Environment, annexes. Unpublished Report, 34 p.

Niell, F.X. & Pazó, J.P. 1978. Incidencia de vertidos industriales en la estructura de poblaciones intermareales. II. Distribución de la biomasa y de la diversidad específica de comunidades de macrófitos de facies rocosa. *Investigaciones Pesqueras*, 42(2): 231–239.

Nielsen, S.N. 1990 Application of exergy in structural–dynamical modelling. *Verhein International Verein Limnology*, 24: 641–645.

Nielsen, S.N. 1992. Strategies for structural dynamics modelling. *Ecological Modelling*, 63: 91–101.

Nielsen, S.N. 1994. Modelling structural dynamic changes in a Danish Shallow Lake. *Ecological Modelling*, 73: 13–30.

Nielsen, S.N. 1995. Optimisation of exergy in a structural dynamic model. *Ecological Modelling*, 77: 111–112.

Niemela, J., Langor, D. & Spence, J.R. 1993. Effects of clear-cut harvesting on boreal ground-bettle assemblages (Coleoptera: Carabidae) in western Canada. *Conservation Biology*, 7: 551–561.

Niencheski, L.F. 1982. Use of *Mytilus galloprovincialis* as pollution indicator in the French Mediterranean coast. Chlorinated organic compounds and heavy metals. *Atlantica*, 5(2): 85–86.

Nilsson, H.C. & Rosenberg, R. 1997. Benthic habitat quality assessment of an oxygen stressed fjord by surface and sediment profile images. *Journal of Marine Sciences*, 11: 249–264.

Nilsson, H.C. & Rosenberg, R. 2000. Succession in marine benthic habitat and fauna in response to oxygen deficiency: analysed by sediment profile-imaging and by grab samples. *Marine Ecology Progress Series*, 197: 139–149.

Niquil, N., Arias-González, J.E., Delesalle, B. & Ulanowicz, R.E. 1999. Characterization of the planktonic food web of Takapoto Atoll lagoon, using network analysis. *Oecologia*, 118: 232–241.

O'Connor, B.D.S., Costolloe, J., Keegan, B.F. & Rhoads, D.C. 1989. The use of REMOTS technology in monitoring coastal enrichment resulting from mariculture. *Marine Pollution Bulletin*, 20: 384–390.

O'Connor, J.S. & Dewling, R.T. 1986. Indices of marine degradation: their utility. *Environmental Management*, 10: 335–343.

Ohlson, M., Söderström, L., Hörnberg, G., Zackrisson, O. & Hermansson, J. 1997. Habitat qualities versus long-term continuity as determinants of biodiversity in boreal old-growth swamp forests. *Biological Conservation*, 81: 221–231.

Orfanidis, S., Panayotidis, P. & Stamatis, N. 2001. Ecological evaluation of transitional and coastal waters: a marine benthic macrophytes-based model. *Mediterranean Marine Science*, 2: 45–65.

Orfanidis, S., Panayotidis, P. & Stamatis, N. 2003. An insight to the ecological evaluation index (EEI). *Ecological Indicators*, 3: 27–33.

OSPAR/MON. 1998. *Report of ad hoc working group on monitoring* (MON), Copenhagen, 23–27 February 1998.

Palmer, A.R. 1994. Fluctuating asymmetry analyses: a primer. In: Markow, T.A. (Ed.), *Developmental Instability: Its Origins and Evolutionary Implications*. Kluwer Academic Publishers, Netherlands, pp. 335–364.

Palmer, A.R. & Strobeck, C. 1986. Fluctuating asymmetry: measurement, analysis, patterns. *Annual Review of Ecology and Systematics*, 17: 391–421.

Pardal, M.A., Cardoso, P.G., Sousa, J.P., Marques, J.C. & Raffaelli, D. 2004. Assessing environmental quality: a novel approach. *Marine Ecology Progress Series*, 267: 1–8.

Pardal, M.A., Marques, J.C., Metelo, I., Lillebø, A.I. & Flindt, M.R. 2000. Impact of eutrophication on the life cycle, population dynamics and production of *Ampithoe valida* (Amphipoda) along an estuarine spatial gradient (Mondego estuary, Portugal). *Marine Ecology Progress Series*, 196: 207–219.

Patrício, J. & Marques, J.C. 2006. Mass balanced models of the food web in three areas along a gradient of eutrophication symptoms in the south arm of the Mondego estuary (Portugal). *Ecological Modelling*, 197: 21–34.

Patrício, J., Neto, J.M., Teixeira, H. & Marques, J.C. 2007. Opportunistic macroalgae metrics for transitional waters. Testing tools to assess ecological quality status in Portugal. *Marine Pollution Bulletin*, 54: 1887–1896.

Patrício, J., Neto, J.M., Teixeira, H., Salas, F. & Marques, J.C. A multi-year comparison of ecological indicators of benthic macrofaunal community condition at a reference and eutrophic site in an estuarine system. *Marine Environmental Research* (submitted).

Patrício, J., Ulanowicz, R., Pardal, M.A. & Marques, J.C. 2004. Ascendency as ecological indcator: A case study on estuarine pulse eutrophication. *Estuarine Coastal and Shelf Science*, 60: 23–35.

Paul, J.F., Scott, K.J., Campbell, D.E., Gentile, J.H., Strobel, C.S., Valente, R.M., Weisberg, S.B., Holland, A.F. & Ranasinghe, J.A. 2001. Developing and applying a benthic index of estuarine condition for the Virginian Biogeographic Province. *Ecological Indicators*, 1: 83–99.

Pearson, T.H. & Rosenberg, R. 1978. Macrobenthic succession in relation to organic enrichment and pollution of the marine environment. *Oceanography and Marine Biology: An Annual Review*, 16: 229–331.

Penna, N., Capellacci, S. & Ricci, F. 2004. The influence of the Po River discharge on phytoplankton bloom dynamics along the coastline of Pesaro (Italy) in the Adriatic Sea. *Marine Pollution Bulletin*, 48: 321–326.

Pérez-Ruzafa, A., 1989. *Estudio ecológico y bionómico de los poblamientos bentónicos del Mar Menor (Murcia, SE de España)*. PhD Thesis, University of Murcia, Spain.

Pérez-Ruzafa, A., Aliaga, V., Barcala, E., García-Charton, J.A., González-Wangúemert, M., Gutiérrez-Ortega, J.M., Marcos, C. & Salas, F. 1994. *Caracterización y valoración del medio marino que constituye el entorno físico y biótico de la central térmica de Escombreras*. Documento científico restringido. Universidad de Murcia.

Pérez-Ruzafa, A., Aliaga, V., Marcos, C. & Salas, F. 1997. *Seguimiento del impacto producido por las jaulas suspendidas para engorde de atunes sobre las comunidades bentónicas del área de Cabo Tiñoso*. Documento científico restringido. Universidad de Múrcia.

Pérez-Ruzafa, A. & Marcos, C. 1992. Colonization rates and dispersal as essential parameters in the confinement theory to explain the structure and horizontal zonation of lagoon benthic assemblages. *Rapport Commission Internationale pour la Mer Mediterranee*, 33: 100.

Pérez-Ruzafa, A., Marcos, C. & Gilbert, J. 2005. The ecology of the Mar Menor coastal lagoon: a fast changing ecosystem under human pressure. In: Gönenc, J.E. & Wolflin, J. (Eds.), *Coastal Lagoons: Ecosystem Processes and Modelling for Sustainable Use and Development*. CRC Press, Boca Raton, pp. 392–422.

Pérez-Ruzafa, A., Ros, J.D., Marcos, C., Ballester, R. & Pérez-Ruzafa, I.M. 1989. Distribution and biomass of the macrophyte beds in a hypersaline coastal lagoon (the Mar Menor, SE Spain), and its recent evolution following major environmental changes. In: Boudoresque, C.F., Meinenesz, A., Fresi, E. & Gravez, V. (Eds.), *International Workshops on Posidonia Beds*. GIS Posidonie Publications, France, Vol. 2, pp. 49–62.

Pérez-Ruzafa, I. 2003. Efecto de la contaminaciónsobre la vegetación submarina y su valor indicador. In: Pérez-Ruzafa, A., Marcos, C., Salas, F. & Zamora, S. (Eds.), *Perspectivas y herramientas en el estudio de la contaminación marina*. Servicio de Publicaciones, Universidad de Murcia. Murcia, pp. 133–147.

Pergent-Martini, C. 1998. *Posidonia oceanica*: a biological indicator of past and present mercury contamination in the Mediterranean sea. *Marine Environmental Research*, 45: 101–111.

Pergent-Martini, C., Leoni, V., Pasqualini, V., Ardizzone, G.D., Balestri, E., Bedini, R., Belluscio, A., Belsher, T., Borg, J., Boudouresque, C.F., Boumaza, S., Bouquegneau, J.M., Buia, M.C., Calvo, S. *et al.* 2005. Descriptors of *Posidonia oceanica* meadows: use and application. *Ecological Indicators*, 5: 213–230.

Persoone, G. & DePauw, N. 1979. Systems of biological indicators for water quality assessment. *In: Biological aspects of freshwater pollution: Proceedings*. Pergamon Press, Oxford, UK, pp. 39–75.

Petrov, A.N. 1990. *Study on ecology of molluscs (the Black Sea bivalves) employing some relevant indices.* PhD Thesis, Sevastopol, 286 p.

Petrov, A.N. & Shadrina, L.A. 1996. The assessment of impact of the Tchernaya river estuary and municipal sewages on state of marine communities in Sevastopol bay (the Black Sea). In: *Estuarine Environments and Biology of Estuarine Species.* Crangon publishers, Gdansk, 161–169 pp.

Philips, D.H. 1977. The use of biological indicator organisms to monitor trace metal pollution in marine and estuarine environments – a review. *Environmental Pollution,* 13: 281–317.

Picard-Berube, M. & Cossa, D. 1983. Teneurs en benzo 3,4 pyréne chz *Mytilus edulis.* L. de léstuarie et du Golfe du Saint-Laurent. *Marine Environmental Research,* 10: 63–71.

Pielou, E.C. 1969. *An Introduction to Mathematical Ecology.* Wiley Interscience, New York, p. 286.

Pinto, R., Patrício, J., Baeta, A., Fath, B.D., Neto, J.M. & Marques, J.C. 2009. Review and evaluation of estuarine biotic índices to assess benthic condition. *Ecological Indicators,* 9: 1–25.

Pires, S. & Muniz, P. 1999. Trophic structure of polychaetes in the Sao Sebastiao Channel (southeastern Brazil). *Marine Biology,* 5: 517–528.

Plafkin, J.L., Barbour, M.T., Porte, K.D., Gross, S.K. & Hughes, R.M. 1989. Rapid bioassessment protocols for use in streams and rivers. Benthic macroinvertebrates and Fish. EPA/444/4-89/001. Office and Water Regulations and Standards, US. Environmental protection Agency, Washington, DC.

Planas, M. & Mora, J. 1987. Estado de conocimiento actual del bentos en zonas organicamente enriquecidas. *Thalassas,* 5(1): 125–134.

Pocklington, P., Scott, D.B. & Schafer, C.T. 1994. Polychaete response to different aquaculture activities. *Proceedings of the 4th International Polychaete Conference,* 162: 511–520.

Price, P.W. 1988. An overview of organismal interactions in ecosystems in evolutionary and ecological time. *Agriculture, Ecosystems and Environment,* 24: 369–377.

Prior, A., Miles, A.C., Sparrow, A.J. & Price, N. 2004. Development of a classification scheme for the marine benthic invertebrate component, Water Framework Directive. Phase I & II – Transitional and coastal waters. R & D Interim Technical Report E1-116/E1-132, 103 p.

Prygiel, J. & Coste, M. 1999. Progress in the use of diatoms for monitoring rivers in France. In: Prygiel, J., Whitton, B.A. & Bukowska, J. (Eds.), *Use of Algae for Monitoring Rivers III.* Agence de l'Eau Artois-Picardie, Douai (France), pp. 165–179.

Prygiel, J. & Coste, M. 2000. Guide méthodologique pour la mise en oeuvre de l'Indice Biologique Diatomées NF T 90–354. Agences de l'Eau – Cemagref-Groupement de Bordeaux. Agences de l'Eau, mars 2000, 134 pp + clés de détermination (90 planches couleurs) + cédérom bilingue français-anglais (Tax'IBD).

Purtauf, T., Dauber, J. & Wolters, V. 2005. The response of carabids to landscape simplification differs between trophic groups. *Oecologia,* 142: 458–464.

Raffaelli, D.G. & Mason, C.F. 1981. Pollution monitoring with meiofauna using the ratio of nematodes to copepods. *Marine Pollution Bulletin,* 12: 158–163.

Ramm, A.E. 1988. The community degradation index: a new method for assessing the deterioration of aquatic habitats. *Water Resources,* 22: 293–301.

Ramm, A.E. 1990. Application of the community degradation index to South African estuaries. *Water Resources,* 24: 383–389.

Raposo, P. 1996. Biología e Ecologia de *Liza ramada* (Risso, 1826) e *Chelon labrosus* (Risso, 1826) (Pisces, Mugilidae). Inter-relações com o ecosistema estuarino. PhD Thesis, University of Lisbon.

Refseth, D. 1980. Ecological analyses of carabid communities – potential use in biological classification for nature conservation. *Biological Conservation*, 17: 131–141.

Regoli, F. & Orlando, E. 1993. *Mytilus galloprovincialis* as a bioindicator of lead pollution: biological variables and cellular responses. In: Sloof, W. & De Kruijf, H. (Eds.), *Proceedings of the Second European Conference on Ecotoxicology*, pp. 1–2.

Reish, D.J. 1993. Effects of metals and organic compounds on survival and bioaccumulation in two species of marine gammaridean amphipod, together with a summary of toxicological research on this group. *Journal of Natural History*, 27: 781–794.

Reish, D.J. & Gerlinger, T.V. 1984. The effects of cadmium, lead, and zinc on survival and reproduction in the polychaetous annelid *Neanthes arenaceodentata* (F. Nereididae). *Proceedings of the First International Polychaete Conference:* 383–389.

Reiss, H. & Kröncke, I. 2005. Seasonal variability of benthic indices: an approach to test the applicability of different indices for ecosystem quality assessment. *Marine Pollution Bulletin*, 50: 1490–1499.

Reizopoulou, S., Thessalou-Legaki, M. & Nicolaidou, A. 1996. Assesment of disturbance in Mediterranean lagoons: an evaluation of methods. *Marine Biology*, 125: 189–197.

Renberg, L., Tarkpea, M. & Sundstroem, G. 1986. The use of the bivalve *Mytilus edulis* as a test organism for bioconcentration studies: II. The bioconcentration of two super(14)C-labelled chlorinated paraffins. *Ecotoxicology and Environmental Safety*, 11(3): 361–372.

Resh, V.H. & Jackson, J.K. 1993. Rapid assessment approaches to biomonitoring using benthic invertebrates, In: Rosenberg, D.M. & Resh, V.H. (Eds.), *Freswater Biomonitoring and Benthic Invertebrates*. Chapman & Hall, N.Y., pp. 194–219.

Resh, V.H. & McElravy, E.P. 1993. Contemporary quantitative approaches to biomonitoring using benthic macroinvertebrates. In: Rosenberg, D.M. & Resh, V.H. (Ed.), *Freshwater Biomonitoring and Benthic Macroinvertebrates*. Chapman & Hall, New York, London, pp. 159–194.

Reynaud, P.A. & Thiolouse, J. 2000. Identification of birds as biological markers along a neotropical urban–rural gradient (Cayenne, French Guiana) using co-inertia analysis. *Journal of Environmental Management*, 59: 121–140.

Rhoads, D.C. & Germano, J.C. 1982. Characterization of organism–sediment relations using sediment profile imaging: an efficient method of Remote Ecological Monitoring of the sea floor (REMOTS™ System). *Marine Ecology Progress Series*, 8: 115–128.

Rhoads, D.C. & Germano, J.C. 1986. Interpreting long-term changes in benthic community structure: a new protocol. *Hydrobiologia*, 142: 291–308.

Richardson, B.J. & Waid, J.S. 1983. Polychlorinated biphenyls (PCBs) in Shellfish from Australian coastal waters. *Ecological Bulletins (Stockholm)*, 35: 511–517.

Riisgard, H.U., Kierboe, T., Molenberg, F., Drabaek, I. & Madsen, P. 1985. Accumulation, elimination and chemical speciation of mercury in the bivalves *Mytilus edulis* and *Macoma balthica. Marine Biology*, 86: 55–62.

Ringwood, A.H., Brouwer, M., Forlin, L. & Andersson, T. 1995. Patterns of metallothionein expression in oyster embryos. *Marine Environmental Research*, 39(1/4): 101–105.

Ritz, D.A., Lewis, M.E. & Shen, M. 1989. Response to organic enrichment of infaunal macrobenthic communities under salmonid seacages. *Marine Biology*, 103: 211–214.

Ritz, D.A., Swain, R. & Elliot, N.G. 1982. Use of the mussel *Mytilus edulis planulatus* (Lamarck) in monitoring heavy metal levels in seawater. *Australian Journal of Marine and Freshwater Research*, 33(3): 491–506.

Roberts, R.D., Gregory, M.G. & Foster, B.A. 1998. Developing an efficient macrofauna monitoring index from an impact study – a dredge spoil example. *Marine Pollution Bulletin*, 36(3): 231–235.

Robinson, G.R., Yurlina, M.E. & Handel, S.N. 1994. A century of change in the Staten Island flora: ecological correlates of species losses and invasions. *Bulletin of the Torrey Botanical Club*, 121: 119–129.

Roesijadi, G. 1994. Metallothionein induction as a measure of response to metal exposure in aquatic animals. *Genetic and Molecular Ecotoxicology*, 102(12): 91–96.

Rogers, S.I., Clarke, K.R. & Reynolds, J.D. 1999. The taxonomic distinctness of coastal bottom-dwelling fish communities of the Northeast Atlantic. *Journal of Animal Ecology*, 68: 769–788.

Romeo, M. & Gnassia-Barelli, M. 1988. *Donax trunculus* and *Venus verrucosa* as bio-indicators of trace metal concentrations in Mauritanian coastal waters. *Marine Biology*, 99(2): 223–227.

Romero, J., Alcoverro, T., Martínez-Crego, B. & Pérez, M. 2005. The seagrass *Posidonia oceanica* as a quality element under the Water Framework Directive: POMI, a multi-variate method to assess ecological status of Catalan coastal waters. Working document of the POMI group, Universitty of Barcelona and Centre d'Estudis Avançats de Blanes-CSIC.

Ros, J.D. & Cardell, M.J. 1991. La diversidad específica y otros descriptores de contaminación orgánica en comunidades bentónicas marinas. *Actas del Symposium sobre Diversidad Biológica.* Centro de Estudios Ramón Areces, Madrid, 219–223 pp.

Ros, J.D., Cardell, M.J., Alva, V., Palacin, C. & Llobet, I. 1990. Comunidades sobre fondos blandos afectados por un aporte masivo de lodos y aguas residuales (litoral frente a Barcelona, Mediterráneo occidental): resultados preliminares. *Bentos*, 6: 407–423.

Rosenberg, D.M. & Resh, V.H. 1993. Introduction to freshwater biomonitoring and benthic macroinvertebrates. In: Rosenberg, D.M. & Resh, V.H. (Eds.), *Freshwater Biomonitoring and Benthic Macroinvertebrates.* Chapman & Hall, New York, London, pp. 1–9.

Rosenberg, R., Blomquist, M., Nilsson, H.C., Cederwall, H. & Dimming, A. 2004. Marine quality assessment by use of benthic species-abundance distribution: a proposed new protocol within the European Union Water framework Directive. *Marine Pollution Bulletin*, 49: 728–739.

Ross, L.T. & Jones, D.A. (Eds.). 1979. *Biological aspects of water quality in Florida.* Technical Series Volume 4, no. 3. Florida Department of Environmental Regulation, Tallahassee.

Rott, E., Van Dam, H., Pfister, P., Pipp, E., Pall, K., Binder, N. & Ortler, K. 1999. *Indikationslisten für Aufwuchsalgen. Teil 2: Trophieindikation, geochemische Reaktion, toxikologische und taxonomische Anmerkungen.* Publ. Wasserwirtschaftskataster, BMfLF, 248 pp.

Ruiz, J.L., Perez, M.R. & Romero, J. 2001. Effects of fish farm loadings on seagrass (*Posidonia oceanica*): distribution, growth and photosynthesis. *Marine Pollution Bulletin*, 42(9): 749–760.

Rygg, B. 1985. Distribution of species along a pollution gradient induced diversity gradients in benthic communities in Norwegian Fjords. *Marine Pollution Bulletin*, 16(12): 469–473.

Rygg, B. 2002. Indicator species index for assessing benthic ecological quality in marine waters of Norway. Norwegian Institute for Water Research, Report no. 40114, 1–32 pp.

Salas, F. 2002. *Valoración y aplicabilidad de los índices y bioindicadores de contaminación orgánica en la gestión del medio marino*. PhD Thesis, University of Murcia, Spain.

Salas, F., Marcos, C., Pérez-Ruzafa, A. & Marques, J.C. 2005. Application of the Exergy Index as ecological indicator along an organic enrichment gradient in the Mar Menor Lagoon (South-Eastern Spain). *Energy*, 30: 2505–2522.

Salas, F., Neto, J.M., Borja, Á. & Marques, J.C. 2004. Evaluation of the applicability of a Marine Biotic Index to characterise the status of estuarine ecosystems: the case of Mondego estuary (Portugal). *Ecological Indicators*, 4: 215–225.

Sanders, H.L. 1968. Marine benthic diversity: a comparative study. *American Naturalist*, 102: 243–282.

Satsmadjis, J. 1982. Analysis of benthic fauna and measurement of pollution. *Reviews in International Oceanography and Medicine*, 66–67:103–107.

Satsmadjis, J. 1985. Comparison of indicators of pollution in the Mediterranean. *Marine Pollution Bulletin*, 16: 395–400.

Scanlan, C.M., Foden, J., Wells, E. & Best, M.A. 2007. The monitoring of opportunistic macroalgal blooms for the water framework directive. *Marine Pollution Bulletin*, 55: 162–171.

Schimmel, S.C., Benyi, S.J. & Strobel, C.J. 1999. An assessment of the ecological condition of Long Island Sound, 1990–1993. *Environmental Monitoring and Assessment*, 56: 27–49.

Schramm, W. 1999. Factors influencing seaweed responses to eutrophication: some results from EU-project EUMAC. *Journal of Applied Phycology*, 11: 69–78.

Shackleford, B. 1988. *Rapid bioassiessments of lotic macroinvertebrate communities: biocriteria development*. Arkansas Deparment of Pollution Control and Ecology, Little Rock, Arkansas.

Shannon, C.E. & Weaver, W. 1963. *The Mathematical Theory of Communication*. University of Illinois Press, Chicago.

Shaw, K.M., Lambshead, P.J.D. & Platt, H.M. 1983. Detection of pollution-induced disturbance in marine benthic assemblages with special reference to nematodes. *Marine Ecology Progress Series*, 11: 195–202.

Simboura, N., Panayotidis, P. & Papathanassiou, E. 2005. A synthesis of the biological quality elements for the implementation of the European Water Framework Directive in the Mediterranean ecoregion: the case of Saronikos Gulf. *Ecological Indicators* 5: 253–266.

Simboura, N., Papathanassiou, E. & Sakellariou, D. 2007. The use of a biotic index (Bentix) in assessing long-term effects of dumping coarse metalliferous waste on soft bottom benthic communities. *Ecological Indicators* 7(1): 164–180.

Simboura, N. & Zenetos, A. 2002. Benthic indicators to use in ecological quality classification of Mediterranean soft bottom marine ecosystems, including a new biotic index. *Mediterranean Marine Science*, 3: 77–111.

Simpson, E.H. 1949. Measurement of diversity. *Nature*, 163: 688.

Smith, R.W., Bergen, M., Weisberg, S.B., Cadien, D., Dalkey, A., Montagne, D., Stull, J.K. & Velarde, R.G. 1998. Southern California Bight Pilot Project: Benthic Response Index for Assessing Infaunal Communities on the Mainland Shelf of Southern California. Southern California Coastal Water Research Project. http://www.sccwrp.org

Smith, R.W., Bergen, M., Weisberg, S.B., Cadien, D., Dalkey, A., Montagne, D., Stull, J.K. & Velarde, R.G. 2001. Benthic response index for assessing infaunal communities on the Mainland shelf of southern California. *Ecological Applications*, 11: 1073–1087.

Somerfield, P.J., Clarke, K.R. & Warwick, R.M. 2003. *Taxonomic Distinctness as an Indicator of Biodiversity and Environmental Health: Analysis of Data and Draft Recommendations.* Plymouth Marine Laboratory, Plymouth.

Somerfield, P.J., Olsgard, F. & Carr, M.R. 1997. A further examination of two new taxonomic distinctness measures. *Marine Ecology Progress Series*, 154: 303–306.

Spies, R.B., Felton, J.S. & Dillard, L. 1984. Hepatic mixed-function oxidases in California Flatfishes are increased in contaminated environments and by oil and PCB ingestion. *Response of Marine Organisms to Pollutants*, 14(1–4): 412–413.

Spitzer, K., Jaros, J., Havelka, J. & Leps, J. 1997. Effects of small-scale disturbance on butterfly communities of an Indochinese montane rainforest. *Biological Conservation*, 80: 9–15.

Stark, J.D. 1985: A macroinvertebrate community index of water quality for stony streams. *Water & Soil miscellaneous publication, 87.* Wellington, New Zealand, National Water and Soil Conservation Authority Wellington. 53 p.

Stark, J.D. 1998. SQMCI: a biotic index for freshwater macroinvertebrates coded-abundance data. *New Zealand Journal of Marine and Freshwater Research*, 32: 55–66.

Stewart, P.L. 1994. Environmental requirements of blue mussel (Mytilus edulis) in eastern Canada and its response to human impacts. *Canadian Technical Report on Fisheries and Aquatic Science*, 3: 1–54.

Stirn, J., Avcin, A.J., Kerzan, I., Marcotte, B.M., Meith-Avcin, N., Vriser, B. & Vukovic, S. 1971. Selected biological methods for assessment of pollution. In: Pearson, E.A. & Defraja, E. (Eds.), *Marine Pollution and Waste Disposal.* Pergamon Press, Oxford, pp. 307–328.

Storelli, M.M & Marcotrigiano, G.O. 2001. Persistent organochlorine residues and toxic evaluation of polychlorinated biphenyls in sharks from the Mediterranean Sea (Italy). *Marine Pollution Bulletin,* 42(12): 1323–1329.

Straškraba, M. 1983. Cybernetic formulation of control in ecosystems. *Ecological Modelling*, 18: 85–98.

Szaro, R.C. 1986. Guild management: an evaluation of avian guilds as a predictive tool. *Environmental Management*, 10: 681–688.

Taghon, G.L., Nowell, A.R. & Jumars, P. 1980. Induction of suspension feeding in spionid Polychaetes by high particulate fluxes. *Science*, 210: 562–564.

Teixeira, H., Neto, J. M., Patrício, J., Veríssimo, H., Pinto, R., Salas, F. & Marques, J.C. Quality assessment of benthic macroinvertebrates under the scope of WFD. P-BAT, the Portuguese Benthic Assessment Tool. *Marine Pollution Bulletin.* (submitted)

Teixeira, H., Salas, F., Borja, Á., Neto, J.M. & Marques, J.C. 2008a. A benthic perspective in assessing the ecological status of estuaries: the case of the Mondego estuary (Portugal). *Ecological Indicators*, 8(4): 404–416.

Teixeira, H., Salas, F., Neto, J.M., Patrício, J., Pinto, R., Veríssimo, H., García-Charton, J.A., Marcos, C., Pérez-Ruzafa, A. & Marques, J.C. 2008b. Ecological indices tracking distinct impacts along disturbance–recovery gradients in a temperate NE Atlantic Estuary – guidance on reference values. *Estuarine Coastal and Shelf Science*, 80(1): 130–140.

Teixeira, H., Salas, F., Pardal, M.A. & Marques, J.C. 2007. Applicability of ecological evaluation tools in estuarine ecosystems: the case of Mondego Estuary (Portugal). *Hydrobiologia*, 587: 101–112.

Tolkamp, H.H. 1985. Biological Assessment of water quality in running water using Macroinvertebrates: a case study for Limburg, The Netherlands. *Water Science and Technology*, 17(17): 867–878.

Uetz, G.W. 1976. Gradient analysis of spider communities in a streamside forest. *Oecologica*, 22: 373–385.

Ulanowicz, R.E. 1980. An hypothesis on the development of natural communities. *Journal of Theoretical Biology*, 85: 23–245.

Ulanowicz, R.E. 1986. *Growth and Development Ecosystems Phenomenology.* Springer-Verlag, New York.

Ulanowicz, R.E. 1997. *Ecology, the Ascendent Perspective.* Columbia University Press, New York, p. 201.

Ulanowicz, R.E. 2000. Ascendency: a measure of ecosystem performance. In: Jørgensen, S.E. & Müller, F. (Eds.), *Handbook of Ecosystem Theories and Management.* Lewis Publishers, Boca Raton, pp. 303–315.

Ulanowicz, R.E. & Norden, J.S. 1990. Symmetrical overhead in flow and networks. *International Journal of Food Science*, 21(2): 429–437.

Ulanowicz, R.E. & Wulff, F. 1991. Comparing ecosystem structures: the Chesapeake Bay and the Baltic Sea. In: Cole, J., Lovett, G. & Findlay, S. (Eds.), *Comparative Analyses of Ecosystems: Patterns, Mechanisms and Theories.* Spring Verlag, New York, pp. 140–166.

UNESCO. 2003. A reference guide on the use of indicators for integrated coastal management. ICAM Dossier 1, IOC Manuals and Guides, 45 p.

USEPA. 2006. National Coastal Assessment, Northeast website.

USEPA. 2007. National Coastal Condition Report III. U.S. Environmental Protection Agency, Washington, DC.

Vaas, P.A. & Jordan, S.J. 1990. Long term trends in abundance indices for 19 species of Chesapeake Bay fishes: reflection of trends in the Bay ecosystem. In: *New perspectives in the Chesapeake System: a Research and Management Partnership.* Proceedings of a Conference.Chesapeake Research Consortium Publ. 137.

Valente, R.M., Rhoads, D.C. & Germano, J.D. 1992. Mapping of benthic enrichment patterns in Narragansett Bay, Rhode Island. *Estuaries*, 15: 1–17.

Van de Bund, W., Cardoso, A.C., Heiskanen, A. & Nõges, P. 2004. Overview of common Intercalibration types. EC-JRC, available at: http://circa.europa.eu/Public/irc/jrc/jrc_eewai/library?l=/intercalibration/typesmanual_2004pdf/_EN_1.0_&a=d in 10/12/2007).

Van der Pijl, L. 1972. *Principles of Dispersal in Higher Plants.* Springer-Verlag, Berlin.

Van Dolah, R.F., Hyland, J.L., Holland, A.F., Rosen, J.S. & Snoots, T.R. 1999. A benthic index of biological integrity for assessing habitat quality in estuaries of the southeastern USA. *Marine Environmental Research*, 48: 269–283.

Varanasi, U., Chan, S. & Clark, R. 1989. *National Benthic Surveillance Project: Pacific coast.* National Ocean Service. Washington.

Verdelhos, T., Neto, J.M., Marques, J.C. & Pardal, M.A. 2005. The effect of eutrophication abatement on the bivalve *Scrobicularia plana*. *Estuarine, Coastal and Shelf Science*, 63: 261–268.

Verlaque, M. 1977. Impact du rejet thermique de Martigues-Ponteau sur le macrophyto-benthos. *Tethys*, 8(1): 19–46.

Viarengo, A. & Canesi, L. 1991. Mussels as biological indicators of pollution. *Aquaculture*, 94: 225–243.

Villalba, A. & Vieitez, J.M. 1985. Estudio de la fauna de anélidos poliquetos del sustrato rocoso intermareal de una zona contaminada de la Ría de Pontevedra (Galicia). *Cahiers De Biologie Marine*, 26: 359–377.

Vincent, C., Heinrich, H., Edwards, A., Nygaard, K. & Haythornthwaite, J. 2002. Guidance on typology, classification and reference conditions for transitional and coastal waters. European Commission, report of CIS WG2.4 (COAST), 119 pp.

Vollenweider, R.A., Giovanardi, F., Montanari, G. & Rinaldi, A. 1998. Characterisation of the trophic conditions of marine coastal waters with special reference to the NW Adriatic Sea: proposal for a trophic scale, turbidity and generalised water quality index. *Environmetrics*, 9: 329–357.

Warwick, R.M. 1986. A new method for detecting pollution effects on marine macrobenthic communities. *Marine Biology*, 92: 557–562.

Warwick, R.M. 1993. Environmental impact studies on marine communities: pragmatical considerations. *Australian Journal of Ecology*, 18: 63–80.

Warwick, R.M. & Clarke, K.R. 1994. Relearning the ABC: taxonomic changes and abundance/biomass relationships in disturbed benthic communities. *Marine Biology*, 118: 739–744.

Warwick, R.M. & Clarke K.R. 1995. New 'biodiversity' measures reveal a decrease in taxonomic distinctness with increasing stress. *Marine Ecology Progress Series*, 129:, 301–305.

Warwick, R.M. & Clarke, K.R. 1998. Taxonomic distinctness and environmental assessment. *Journal of Applied Ecology*, 35: 532–543.

Warwick, R.M. & Clarke, K.R. 2001. Practical measures of marine biodiversity based on relatedness of species. *Oceanography and Marine Biology: An Annual Review*, 39: 207–231.

Weisberg, S.B., Ranasinghe, J.A., Dauer, D.M., Schaffner, L.C., Díaz, R.J & Frithsen, J.B. 1997. An estuarine benthic index of biotic integrity (B-IBI) for the Chesapeake Bay. *Estuaries*, 20: 149–158.

Wells, E., Wood, P., Wilkinson, M. & Scanlan, C. 2007. The use of macroalgal species richness and composition on intertidal rocky seashores in the assessment of ecologcial quality under the European Water Framework Directive. *Marine Pollution Bulletin*, 55: 151–161.

Weston, D.P. 1990. Quantitative examination of macrobenthic community changes along an organic enrichment gradient. *Marine Ecology Progress Series*, 61(3): 233–244.

Wetzel, R.G. 1983. *Limnology*. Saunders College Publishing, New York. p. 858.

WFD, 2000/60/CE. Directive 2000/60/EC of the European Parliament and of the Council establishing a framework for community action in the field of Water Policy. European Communities 43: 1–72.

WGBEC. 2002. *Report of the Working Group on Biological effects of contaminants*. Murcia, 11–15 March 2002. ICES CM 2002/E: 02 Ref. ACME.

Wieser, W. 1953. Die Beziehung zwischen Mundhö¨hlengestalt, Erna¨hrungsweise und Vorkommen bei freilebenden marinen Nematoden. *Ark Zool*, 4:439–484

Wildish, D.J., Hargrave, B.T., MacLeod, C. & Crawforl, C. 2003. Detection of organic enrichment near finfish net-pens by sediment profile imaging at SCUBA-accessible depths. *Journal of Experimental Marine Biology and Ecology*, 285–286: 403–413.

Wilsey, B.J., Haukioja, E., Koricheva, J. & Sulkinoja, M. 1998. Leaf fluctuating asymmetry increases with hybridization and elevation in tree-line birches. *Ecology*, 79: 2092–2099.

Wilson, J.G. 2003. Evaluation of estuarine quality status at system level using the Biological Quality Index and the Pollution Load Index. Biology and Environment: Proceedings of the Royal Irish Academy, Vol. 103B N° 2: 49–57.

Wilson, J.G., Ducrotoy, J.P., Desprez, M. & Elkaim, B. 1987. Application of two estuary quality indices to the central and western channel: status of Somme and Seine estuaries (France). *Vie Milieu*, 37: 1–11.

Wither, A. 2003. Guidance for sites potentially impacted by algal mats (green seaweed). EC Habitats Directive Technical Advisory Group report WQTAG07c.

Woodiwiss, F.S. 1964: The biological system of stream classification used by the Trent River Board. *Chemistry and Industry*, 83: 443–447.

Wooton, M. & Lye, A.K. 1982. Metal levels in the mussel *Mytilus edulis* collected from estuaries in south eastern Australia. *Australian Journal of Marine and Freshwater Research*, 32(2): 363–367.

Word, J.Q. 1979. The infaunal trophic index. *California Coastal Water Research Project Annual Report*, 19–39 pp.

Word, J.Q. 1980. *The infaunal trophic index. The 1980 Annual Report, Southern California Coastal Research Project*. Long Beach, CA, p. 19–39.

Word, J.Q. 1990. *The infaunal trophic index a functional approach to benthic community analyses*. PhD Thesis. University of Washington, 297 p.

Yokoyama, H. 1997. Effects of fish farming on macroinvertebrates. Comparison of three localities suffering from hypoxia. *UJNR Technical Report*, 24: 17–24.

Zabala, K., Romero, A. & Ibáñez, M. 1983. La contaminación marina en Guipuzcoa: I. Estudio de los indicadores biológicos de contaminación en los sedimentos de la Ría de Pasajes. *Lurralde*, 3: 177–189.

Zelinka, M. & Marvan, P. 1961. Zur Präzisirung der biologischen Klassifikation der Reinheit fliessender Gewässer. *Archiv für Hydrobiologie*, 57: 389–407.

Zenetos, A., Chadjianestis, I., Lantzouni, M., Simboura, M., Sklivagou, E. & Arvanitakis, G. 2004. The Eurobulker oil spill: midterm changes of some ecosystem indicators. *Marine Pollution Bulletin*, 48(1/2): 121–131.

Zettler, M.L., Schiedek, D. & Bobertz, B., 2007. Benthic biodiversity indices versus salinity gradient in the southern Baltic Sea. *Marine Pollution Bulletin*, 55: 258–270.

Zvereva, E.L., Kozlov, M.V., Niemelä, P. & Haukioja, E. 1997. Delayed induced resistance and increase in leaf fluctuating asymmetry as responses of *Salix borealis* to insect herbivory. *Oecologia*, 109: 368–373.

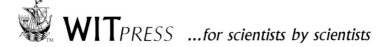

Disaster Management and Human Health Risk

Reducing Risk, Improving Outcomes

Edited by: **K DUNCAN**, University of Toronto, Canada and **C.A. BREBBIA**, Wessex Institute of Technology, UK

Recently, there has been a disturbing increase in the number of natural disasters affecting millions of people, destroying property and resulting in loss of human life. These events include major flooding, hurricanes, earthquakes and many others.

Today the world faces unparalleled threats from human-made disasters that can be attributed to failure of industrial and energy installations as well as to terrorism. Added to this is the unparalleled threat of emerging and re-emerging diseases, with scientists predicting events such as an influenza pandemic.

Containing papers from the First International Conference on Disaster Management and Human Health Risk, Reducing Risk and Improving Outcomes on the following topics: Global Risks and Health; Chemical Emergencies; Extreme Weather Events; Food and Water Safety; Natural Disasters; Pandemics and Biological Threats; Radiation Emergencies; Terrorism; Offshore Disasters; Remote Areas Response; Emergency Preparedness and Planning; Risk Mitigation; Surveillance and Early Warning Systems; Disaster Resilient Communities; Disaster Epidemiology and Assessment; Disaster Mental Health; Business Continuity; Human Health Economics; Recent Incidents and Outbreaks; Public Health Preparedness.

WIT Transactions on The Built Environment, Vol 110
ISBN: 978-1-84564-202-0   2009   apx 400 pp
apx £132.00/US\$264.00/€175.00
eISBN: 978-1-84564-379-9

WITPRESS ...for scientists by scientists

FLOOD RECOVERY,
INNOVATION AND
RESPONSE
EDITED BY
D. Proverbs, C.A Brebbia and
E. Penning-Rowsell

Flood Recovery, Innovation and Response

Edited by: **D. PROVERBS**, *University of Wolverhampton,*
UK, **C.A. BREBBIA**, *Wessex Institute of Technology, UK and*
E. PENNING-ROWSELL, *University of Middlesex, UK*

Recent catastrophes, from the 2004 Indian Ocean
tsunami and the ravaging of New Orleans in the
aftermath of Hurricane Katrina, to the 2007 floods in
Bangladesh, England, and Texas, have made the world
very aware of the need for better management of the
response to flooding and of the rehabilitation of
damaged areas. This book contains papers originally
presented at the First International Conference on Flood Recovery Innovation
and Response (FRIAR), held in London, UK, which brought together
academics, practitioners, and government officials to share information on
the state of the art in this field.

The conference papers address one of seven main themes: Flood Risk
Management Vulnerability Assessment and Modelling; Flood Risk
Management; Flood Defence Methods; Rehabilitation and Restoration; Urban
Flood Management – Innovation in Building Resilience; Coping Strategies;
Financial and Insurance Issues.

WIT Transactions on Ecology and the Environment, Vol 118
ISBN: 978-1-84564-132-0   2008   368pp
£121.00/US$242.00/€157.00
eISBN: 978-1-84564-331-7

WITPress
Ashurst Lodge, Ashurst, Southampton,
SO40 7AA, UK.
Tel: 44 (0) 238 029 3223
Fax: 44 (0) 238 029 2853
E-Mail: witpress@witpress.com

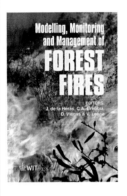

Modelling, Monitoring and Management of Forest Fires

Edited by: **J. de las HERAS**, *Universidad de Castilla La Mancha, Spain,* **C.A. BREBBIA**, *Wessex Institute of Technology, UK,* **D. VIEGAS**, *University of Coimbra, Portugal and* **V. LEONE**, *Universita della Basilicata, Italy*

Forest fires are very complex phenomena which, under the right physical conditions, can rapidly devastate large areas, as demonstrated by recent events. There is also widespread awareness that the risk may increase as a result of climate changes.

Different approaches are required for controlling fires in rural areas as opposed to urban environments, since the approaches strongly depend on the physical characteristics of the surrounding region. They are also functions of weather conditions, regional activities and forest type, as well as social and economic factors.

Containing papers presented at the First International Conference on Modelling, Monitoring and Management of Forest Fires, this book addresses all the aspects of forest fires, from fire propagation in different scenarios to the optimum strategies for fire-fighting. It also covers issues related to economic, ecological, social and health effects.

Featured topics include: Computational Methods and Experiments; Fire Mitigation Models; Decision Support Systems; Laboratory and Field Experiments to Assess Fire Propagation Models; Monitoring Systems; Shrub and Peat Fire Danger Rating; Wildlfe Modelling; Risk and Vulnerability Assessment; Environmental Impact; Air Pollution and Health Risk; Case Studies.

WIT Transactions on Ecology and the Environment, Vol 119
ISBN: 978-1-84564-141-2   2008   432pp
£142.00/US$284.00/€184.50
eISBN: 978-1-84564-341-6

Ecosystems and Sustainable Development VII

Edited by: **E. TIEZZI**, *University of Siena, Italy and* **C.A. BREBBIA**, *Wessex Institute of Technology UK*

ECOSUD is a challenge for the creation of a new science in line with Prigogine's statement that "at all levels we observe events associated with the emergence of novelties and narrative elements, which we may associate with the creative power of nature". This is not only a platform to present novel research related to ecological problems from all over the world; it also gives opportunities for new emergent ideas in science arising from the cross fertilization of different disciplines, including mathematical models and eco-informatics, evolutionary thermodynamics and biodiversity, structures in ecosystems modelling and landscapes to mention but a few.

The Seventh International Conference in the well-established series on Ecosystems and Sustainable Development contains papers on the following topics: Thermodynamics and Ecology; Sustainability Indicators; Mathematical and System Modelling; Ecosystems Modelling; Biodiversity; Sustainability Development Studies; Conservation and Management of Ecological Areas; Socio-Economic Factors; Energy Conservation and Generation; Environmental and Ecological Policies; Environmental Management; Environmental Risk; Natural Resources Management; Recovery of Damaged Areas; Biological Aspects; Complexity; Remote Sensing; Landscapes and Forestation Issues; Soil and Agricultural Issues; Water Resources; Sustainable Waste Managements; Air Pollution and its Effects on Ecosystems.

WIT Transactions on Ecology and the Environment, Vol 122
ISBN: 978-1-84564-194-8   2009   apx 800pp
apx £265.00/US$530.00/€349.00
eISBN: 978-1-84564-371-3

WIT eLibrary

Home of the Transactions of the Wessex Institute, the WIT electronic-library provides the international scientific community with immediate and permanent access to individual papers presented at WIT conferences. Visitors to the WIT eLibrary can freely browse and search abstracts of all papers in the collection before progressing to download their full text.

Visit the WIT eLibrary at
http://library.witpress.com